VGM Opportunities Series

OPPORTUNITIES IN
HEATING, VENTILATION, AIR-CONDITIONING, AND REFRIGERATION CAREERS

Richard Budzik

Foreword by
Frank M. Coda
Executive Director
American Society of Heating, Refrigerating,
and Air-Conditioning Engineers, Inc.

VGM Career Horizons
a division of *NTC Publishing Group*
Lincolnwood, Illinois USA

Library of Congress Cataloging-in-Publication Data
Budzik, Richard S.
    Opportunities in heating, ventilation, air-conditioning, and,
    refrigeration careers / Richard Budzik ; foreword by Frank M. Coda.
        p. cm. — (VGM opportunities series)
    Includes bibliographical references.
    ISBN 0-8442-4589-5 (hard). — ISBN 0-8442-4590-9 (soft)
    1. Heating—Vocational guidance.   2. Ventilation—Vocational
guidance.   3. Air conditioning—Vocational guidance   I. Title.
II. Series
TH7015.B83   1995
697'.0023—dc20   95-648
                                                        CIP

Published by VGM Career Horizons, a division of NTC Publishing Group
4255 West Touhy Avenue
Lincolnwood (Chicago), Illinois 60646-1975, U.S.A.
© 1996 by NTC Publishing Group. All rights reserved.
No part of this book may be reproduced, stored in a retrieval
system, or transmitted in any form or by any means,
electronic, mechanical, photocopying, recording or otherwise,
without the prior permission of NTC Publishing Group.
Manufactured in the United States of America.

5 6 7 8 9 VP 9 8 7 6 5 4 3 2 1

CONTENTS

ABOUT THE AUTHOR

Since 1964 Richard Budzik has been a teacher at Prosser Vocational School in Chicago. He has trained hundreds of students, apprentices, and mechanics in sheet metal work. He conducts both day and evening classes. The day sessions are for high school vocational students readying themselves to enter the trade when they graduate. At night he holds classes for apprentices and also for mechanics wanting to brush-up their skills, advance themselves, or simply keep pace with a changing industry.

Mr. Budzik has served periodically as a curriculum consultant for the Chicago public school system, and he has conducted company training programs for several industrial firms. He has authored a total of twenty-eight books on various phases of sheet metal work, making him one of the most published, if not the most published, author in this field.

He was formerly technical editor of *American Artisan* and currently is "Shop and Jobs Tips" editor of *HAC,* a heating, air-conditioning, and sheet metal monthly magazine. He writes monthly articles concerning problems facing the sheet metal contractor today. These articles cover many aspects of design, fabrication, and installation of residential, commercial, and industrial duct run systems.

FOREWORD

The heating, ventilating, air-conditioning, and refrigeration (HVACR) industry is a challenging, rewarding, and diverse one. By taking the time to read this book, you will become aware of the many specialized careers available.

Career opportunities exist for people with all levels of education. High school graduates or those with doctorate degrees can find a niche in this field.

Because people expect a comfortable environment in their homes and where they work, shop, or go to school, there has been a growing demand for HVACR designers, installers, and service personnel.

But career opportunities in this exciting field extend far beyond comfort control. The industry satisfies the needs of manufacturers by designing systems for food processing, storing, and shipping; for conditioning indoor environments for the manufacture of microprocessor chips for computers; and even for maintaining sterile environments for surgical suites, to name just a few examples. Also, other opportunities exist for minimizing energy usage in buildings or conducting research that will improve the public's quality of life.

In addition, the industry is responsible for solving issues that have worldwide importance. Managing indoor air quality and designing and operating environmentally safe mechanical systems that maintain the integrity of the ozone layer are two critical issues on which the industry is currently focused. These issues demand solutions, and the solutions are being provided by the industry's men and women.

As you can see, the work of one individual affects the lives of many. When that responsibility is coupled with a range of career opportunities—from owning a business to managing a company's division—you can see how personal satisfaction is derived from a career in this industry.

You have taken that critical first step to learn more about the heating, ventilating, air-conditioning, and refrigeration field. The time invested now will bring future benefits, and your work in the industry will be for the good of many people.

Good luck in your career selection!

Frank M. Coda
Executive Director
American Society of Heating, Refrigerating
and Air-Conditioning Engineers, Inc.
(ASHRAE)

INTRODUCTION

The heating, ventilation, air-conditioning, and refrigeration industry continues to grow at a rate equal to, if not faster than, our population. This is true due to the following facts:

- As our standard of living continues to rise, people demand greater comforts regarding the air in their homes and places of work.
- New and exciting energy-saving innovations are continually being developed.
- The applications of solar energy are growing.

Demand for these imaginative, new developments is increasing every day. Rapidly expanding service and new technology in this area constantly provide new jobs. Many opportunities are waiting for people trained in this field. The heating, ventilation, air-conditioning, and refrigeration industry offers essential services and a great variety of career opportunities, job security, and mobility.

This industry provides the technology to control the environment in any enclosed area, from a home to a space capsule. Any desired temperature can be maintained. Humidity can be increased or decreased. The air can be filtered and cleaned of pollutants. This ability to install and maintain special environments for people, products, and perishables is essential to our lives today.

Opportunities unlimited? This rapidly expanding industry with its new technology not only provides new job opportunities but creates numerous possibilities for advancement. The industry encourages personal development. In many cases it subsidizes continuing education. Many people who enter this field find that they can move into other professions within the industry such as sales representative, estimator, drafter, designer, specification writer, field servicer, lab technician, and wholesale operator.

How Do You Get Started?

Think about yourself—your own interest, your talents, how you like to work. Do you like to work alone, indoors, or outdoors? Do you like to work on a team, travel, deal with many people? Do you like to create with your hands as well as your mind? Would you like to manage your own business? Knowing a little bit about yourself can help you decide what type of career to pursue.

If you haven't already done so, be sure to get the knowledge you need to hold a good job in the industry. A high school education is essential with particular emphasis on mathematics, physics, blueprint reading, and metal shop. Vocational-technical schools, private trade schools, and junior colleges offer programs in heating, ventilation, air-conditioning, and refrigeration. Contact your local schools or state department of education for lists of available courses.

Read and study the information in this book. Also talk to school counselors or to local contractors who are a part of the industry in your area.

CHAPTER 1

A VITAL AND EVER-EXPANDING INDUSTRY

The range of opportunities awaiting you in the heating, ventilation, air-conditioning, and refrigeration (HVACR)* industry today is as wide as your abilities and interests and will carry you as far as the amount of time and work you are willing to invest in your future. Skilled workers in this industry are vital to our ever-expanding American industry. Being a person with these much needed and wanted vocational skills, you will be assured of employment and good wages in all parts of the country.

As is true with any vocational trade, you work with your hands and your head, but the "head" part is more significant in this industry than it is in many others. Therefore, learning and knowledge beyond on-the-job training are important here; formal learning is readily available nearly anywhere, and self-directed materials are also convenient, as will be discussed later.

JOB CATEGORIES

The jobs in the HVACR industry are nearly unlimited in variety and opportunity. The one you select will depend primarily upon

---

*Today, the abbreviation HVACR is commonly used by the industry, associations, and trade journals when referring to all facets of work involving "heating, ventilation, air-conditioning, and refrigeration."

your specific interests and desires. However, there are definite categories into which these numerous jobs fit, and they are briefly outlined here. Later, a separate section is devoted to each category. As you will notice when glancing through the various chapters, some of the specific job classifications are required in more than one general category. The general categories include:

- air-conditioning and refrigeration installation (residential and commercial)
- air-conditioning and refrigeration servicing (residential and commercial)
- maintenance and operations
- marine refrigeration
- cold storage and institutional refrigeration
- automotive air-conditioning and transport refrigeration
- sales
- sheet metal ductwork trade
- engineering technicians
- business ownership

WHAT IS HVACR?

In the last thirty years, a popular use of the term *air-conditioning* has evolved to imply merely cooling the air around us when it is too warm. The term *air conditioner* has been used to refer to the unit that performs this function. In some areas, the term *cooling* has become commonly used in place of air-conditioning.

Air-conditioning engineers use a broader explanation. They define air-conditioning as the process of controlling the temperature, humidity, cleanliness, and distribution of air.[1] So you can readily

---

[1]"A Career in Air Conditioning" pamphlet published by Air-Conditioning and Refrigeration Institute, 4301 N. Fairfax Dr., Ste. 425, Arlington, VA 22203.

see how vital the role of the air-conditioning industry has become to our daily lives. It is important for human comfort at work and leisure in homes, schools, offices, and shops. It is vital for many manufacturing processes as they have evolved today, ranging from current uses of plastics to antibiotics. It certainly is important in the transportation of people and products.

Without air-conditioning computers could not function and jet planes could not carry passengers. Other problems would continue, such as summer heat waves causing widespread illnesses and valuable historic documents continuing to deteriorate and fade at accelerated rates as before we had air-conditioning.

The HVACR industry is a young and vital one, compared to most other trades. And there are many challenges ahead. More efficient equipment and systems will continue to be developed to better condition the air in homes, schools, offices, factories, and shopping centers. At the same time, a new emphasis is placed on conserving energy. The use of solar and nuclear energy to cool buildings will be further explored. Advances in air pollution control will possibly come from the HVACR industry.

Another term frequently used in relation to the complete meaning of conditioning the air is "environmental system."[2] However, we will continue to use the more commonly understood abbreviation, HVACR.

Refrigeration provides the preservation of foods and other perishables, such as many medicines. As with air-conditioning, it is much more than a matter of regulating temperature. It also requires air circulation and humidity control as significant factors in any efficient refrigeration system.

Both air-conditioning and refrigeration depend on the same physical principles and utilize much of the same highly special-

[2]"Careers in Creating Environments" pamphlet published by National Air Conditioning Contractors of America, 1513 16th St. NW, Washington, DC 20036.

ized equipment. The primary difference is in the temperature the equipment is designed to produce. For example, to create indoor comfort conditions an air conditioner generally lowers temperatures by 15–20° F, whereas temperature drops of 85–200° F are commonly produced with refrigeration equipment.

HVACR FOR COMFORT AND HEALTH

Of all the conditions that help make up our physical environment, the ones that affect us most are those relating to climate. We frequently talk about how we feel and the weather. Scientists speak of the day when we will be able to manipulate our solar system to create a completely different pattern of climate and weather; that day, however, certainly seems remote or even impossible to us now. In the meantime and more realistically, the climatic conditions of our environment will remain primarily as they are, with wonderful modifications inside buildings, thanks to our current mechanical systems of heating, ventilation, air-conditioning, and refrigeration.

An HVACR system, whether small and simple or large and complex, consists primarily of a combination of motors, pulleys, compressors, valves, coils, piping, electrical wiring, automatic controls, ducts, and other devices. These devices help to control, in a reliable and efficient manner, the temperature, humidity, purity, and circulation of air within a specific small or large enclosed space. For us, it means primarily making our homes and places of work as comfortable and healthy as possible. It plays a big part in determining:

- how we feel,
- how efficiently we work,
- what food we eat, and
- what medications we take, due to the first three factors.

Creating conditions in which people are comfortable includes regulating temperature, humidity, air filtration, air distribution, and other technical functions.

WHAT DOES HVACR INVOLVE?

Primarily, people who work in the HVACR industry have a hand in designing, planning, making, installing, maintaining, or servicing the systems that provide the following air conditions:

Temperature control is provided by heating, refrigeration, and air-conditioning equipment. These range from small window units to huge central plant systems like those needed in cold storage warehouses, hospitals, and other buildings where large-scale heating and/or cooling is required.

Humidity or moisture removal is accomplished automatically as the air is cooled, so the air can actually absorb moisture from the human bodies, which aids in creating more comfortable conditions for us.

Purity of the air is accomplished by mechanically and electronically filtering out many air pollutants including dust, pollen, and cigarette smoke. Briefly, electronic filters put an electrical charge on solid particles (even minute in size) and then attract them, as the air passes through, to a screen that has an opposite charge.

Ventilation is accomplished by blowers and fans, carefully engineered as supply and return ducts. Ventilation adds fresh air from outdoors to provide oxygen we consume, in addition to recirculating cooled air.

Because windows and doors can be kept closed, an added bonus is the great reduction of outside noises and distractions.

These many advantages have made air-conditioning, which was a luxury not too many years ago, become nearly a necessity today. And installation and maintenance of all this equipment in homes and commercial and industrial buildings increases the need for skilled workers every year. This demand appears certain to increase. This also applies to installing and servicing marine, automotive, and other mobile systems.

REQUIREMENTS FOR SUCCESS

Before going into the detailed chapters regarding the various facets of jobs in HVACR, we will consider general factors necessary for success in the industry.

The most important personality trait is a genuine liking for working with tools on a variety of jobs requiring background experience and knowledge. If you possess an average intelligence and manual dexterity, you can sincerely consider the jobs described in detail in later chapters. You must be able to tolerate a variety of relatively undesirable working conditions such as dirt, grease, noise, and safety hazards. You must, naturally, be in good physical shape, including a sound body, good eyesight, and good stamina. On many jobs you have to stand and move around a major portion of the time and lift somewhat heavy tools and equipment.

Some of the jobs require that candidates complete a major portion of their education before beginning the job; others offer either on-the-job training or a combination of on-the-job training and correlated classes, such as an apprenticeship program.

If you have any handicap, consult a person in the industry to help you determine whether it would seriously affect your ability to handle any of these jobs.

EARNINGS

Naturally, people with these skills and knowledge deserve the respect of other craftsworkers and the relatively high pay they receive.

However, the specific salary depends largely on the individual's education, experience, skill, specific job classification, and also job location. Most people also receive health and accident insurance, pension programs, and paid vacations. Specifics regarding salaries are detailed in the chapters describing each area of work within air-conditioning and refrigeration.

SECURITY AND OPPORTUNITY FOR GROWTH

The HVACR industry provides one of our nation's most stable occupations, partly due to the shortage of trained personnel and due to the increasing uses of these types of equipment. The demand for these services is nationwide.

Jobs are available at every level of skill and knowledge. After becoming a mechanic through training (education and experience), you may progress by working actively in the field or in an office or laboratory position.

There are more opportunities for growth in this industry than there are in many other trades. As technology changes, the need grows for skilled people willing to continue learning and keep pace with new developments in HVACR.

GETTING STARTED

As is true in nearly any skilled trade today, high school graduation is generally essential. Specific courses that can help include mathematics, sciences, and shop. Then there are several avenues you can follow, partly depending on the specific job classification in which you are interested. The remainder of this career guide helps you determine which of these alternatives is best for you.

A career video (available to schools) and a sixteen-page booklet (available to individuals) entitled *Your Pipeline to Hot Careers and Cold Cash* was developed by and can be obtained from the following associations:

National Association of Plumbing-Heating-Cooling Contractors
(NAPHCC)
180 South Washington Street
Box 6808
Falls Church, VA   22046–1148

Mechanical Contractors Association of America (MCAA)
1385 Piccard Drive
Rockville, MD   20850–4329

CHAPTER 2

GETTING THE RIGHT EDUCATION

When you attempt to evaluate the educational requirements that qualify a person for a heating, ventilation, air-conditioning, and refrigeration (HVACR) career, the task is very difficult. Why is this so? First of all, jobs in refrigeration and air-conditioning sales, installation, maintenance, service, and operation are found in every segment of commerce, industry, and home ownership. They range from the semiskilled handyperson who performs the simplest operational and maintenance tasks, to the plant superintendent who is responsible for the operation and maintenance of mechanical systems that cost several million dollars to install. Consider also the responsibility for the safety and comfort of hundreds of men and women who occupy the buildings under the care of these workers.

The second obstacle for the person seeking the necessary basic and technical education to qualify for a job in HVACR is the fact that this industry is comprised of many specialized branches. In fact, the field is so broad that one person cannot be qualified in all its aspects. For this reason, the ambitious person seeking a career in this field should first acquire a basic education that will form a solid foundation for the more specific technical education needed to qualify for a good job.

The third obstacle is that a comprehensive investigation of the people in many of these jobs in the industry revealed conflicting views on the subject of basic and technical education requirements. However, most of those interviewed agreed that it would be wise for a person to start preparing for such a career while still in high school, if possible. Then, one should continue the education through a college-level course in refrigeration and air-conditioning technology that would result in an associate in science degree (a two-year course if attending full time). This level of training is available in many public and private trade-technical colleges and local junior colleges.

Getting the most from any education requires foresight and planning. It is important to seek the type of training that will provide you with the best opportunity to develop a solid base on which to cultivate your talents and interests.

Almost any shop or laboratory course offered in junior or senior high school will be beneficial. These courses can acquaint you with the correct use of basic hand tools and basic shop equipment and will help you learn good work practices. One feature of this training is that greater emphasis is placed on developing attitudes and stimulating interest than on specific skill improvement. You are able to explore your desires and evaluate your aptitude for the various types of work. After the basic courses, try to take courses that are more in-depth in order to develop fundamental skills and broaden your foundation for future growth and specialization.

An on-the-job training program with apprenticeship or other classes to attend is another good way to enter the field, but this type of program is more difficult to locate than are school-related courses.

INDUSTRIAL ARTS COURSES IN HIGH SCHOOL

A basic purpose of industrial arts in junior and senior high schools is to acquaint students with the industrial environment in which they live and will work. *Industrial arts* refers to the study of industrial tools, material processes, products, and occupations in shops, laboratories, and drafting rooms. The more specific training objectives involved in an industrial arts program include the following:

- promote safe work habits and proper attitudes toward fellow workers
- develop efficient work habits
- develop an understanding of industrial materials, processes, and products
- teach a wide variety of skills in the use of tools and machines
- teach clear analytical reasoning when performing mechanical and constructive tasks
- foster favorable attitudes toward group activities in constructive projects

Industrial arts education is an element of general education. Although the specific design of an industrial arts program is not directed toward a specific vocational goal, these programs often help inspire students who have a desire to pursue a mechanical career. Meaningful industrial arts curriculums can provide an excellent background for young people who intend to pursue a vocational education.

VOCATIONAL AND INDUSTRIAL EDUCATION

Vocational education is a specialized program designed to provide instruction leading to some degree of occupational competency. But this term refers broadly to all forms of education and training that are designed to prepare people for some type of successful employment. *Vocational-industrial education* includes trade training and training for other industrial occupations, which is our interest here.

Like industrial arts, vocational-industrial education is intended to complement rather than compete with general education. However, unlike industrial arts education, it places major emphasis on developing a specific salable skill and knowledge. Vocational-industrial education curriculums prepare students for specific occupations, provide industry with trained workers, and provide training and retraining for adults who are unemployed or underemployed.

Recently there has been a trend toward changing the level at which vocational education is being taught. The trend today is toward teaching vocations after high school. This is contrary to past procedure, when many vocational institutions operated in high schools. This is probably due to the upward movement of the general education level in the United States and the advancement of technology and automation; so the upgrading of vocational-industrial education follows the same pattern. Today, many find that trade and technical institutes, both public and private, offer excellent facilities and programs. They provide students with the opportunity to pursue occupational training relevant to their needs and desires after completion of high school.

In larger cities, some vocational-industrial education at the high school level is still offered. Cooperative education is one type of vocational education aimed at easing the school-to-work transition problem. In this type of program, a formal relationship is established between the high school and employers in the community. Students who participate in this program are allowed to divide their time between school and work. Their high school curriculum serves to broaden their scope of general education and classroom occupational training, while their attendance at a correlated part-time job affords them the opportunity to receive on-the-job occupational training. Federally aided vocational education programs in some cities also serve to ease the school-to-work transition by providing occupational training oriented to current job market needs.

APPRENTICESHIP TRAINING

The individual's degree of skill determines the quality and complexity of work he or she can do. To develop skills fully one must receive training, and becoming an apprentice is an excellent way to obtain this training.

An apprenticeship is a recognized formal method of learning a skilled trade. For example, in sheet metal ductwork, the vast majority of journeymen have acquired their knowledge and skills through apprenticeship programs. Sheet Metal Workers International Association recognized the need for pertinent apprenticeship training standards for the industry and formed a National Joint Apprenticeship and Training Committee. This committee was assigned the responsibility of formulating training standards in cooperation with what is now the Bureau of Apprenticeship and Training, U.S. Department of Labor. Because of the changes in any industry, it is necessary to review and revise these standards from time to time. Similar efforts have been accomplished in Canada.

The effectiveness of a training program depends, in large measure, on the techiques of training involved. Apprenticeship training combines on-the-job training with related classroom instruction, thus reinforcing the learning processes. In current practice, it provides a job for the trainee while providing an opportunity to acquire knowledge and develop skills. Most apprenticeship training programs are based on a four-year curriculum.

The role of the Bureau of Apprenticeship and Training, a part of the U.S. Department of Labor's Employment and Training Administration, is to assist labor and management in the development, expansion, and improvement of apprenticeship and training programs. The bureau works closely with state apprenticeship agencies, trade and industrial education institutions, and labor and management.

WHAT DO YOU NEED TO KNOW?

1. *Safety.* Safety is probably the single most important factor in the operation and care of any heating, ventilating, refrigerating, and air-conditioning plant or other mechanical system used in modern buildings. One of the most important elements involved in safety is good housekeeping. Every mechanic, technician, operator, and plant engineer is responsible for the maintenance of clean, orderly, and safe workshops, storage areas, and machinery rooms. These workers must understand the action and reaction of refrigerants, industrial gases, and other chemicals when they are exposed to extremes of temperature and pressure such as those that might result from fire.

2. *Arithmetic or vocational mathematics.* Everyone is in agreement on the need for a good working knowledge of arithmetic and vocational mathematics. This knowledge should include the use of formulas and constants that are commonly used in heat-load calculations, areas of surface, contents of tanks, insulating factors, and other elements used in cost estimating in the refrigeration business, plus the simpler applications of algebra and geometry.

3. *Applied physics and chemistry.* Every technician, mechanic, operating engineer, and plant engineer in the refrigeration and air-conditioning industry should have some knowledge of the chemicals used for water softening, cleaning compounds, brine and eutectic solutions, oils, greases, and other substances commonly used in the operation and care of heating, ventilating, refrigeration, air-conditioning, and other mechanical systems. In addition, some knowledge of the properties of refrigerants, measurement of heat, fundamentals of electricity, magnetism, and strength of materials is helpful.

4. *Records, charts, day books, and logs.* Every technician, mechanic, operating engineer, and plant engineer is responsible for keeping charts, records, and logs of plant operation, repairs, maintenance, and departmental activities. These records can include a day book to record tasks assigned and accomplished, service calls, maintenance, and department needs. The operating engineers must keep a regular log of plant temperatures and pressures, humidity readings, and fuel consumption. The chief engineer must have knowledge and records of office procedures, employee-employer relationships, employee benefits, time keeping, shift schedules, hiring and firing policies, and elements of effective supervision.

5. *Parts, supplies, tools, and fuel.* On most jobs the technicians, mechanics, and plant engineers are responsible for ordering parts, supplies, tools, fuel, and replacement equipment for all mechanical systems in the buildings under their care. They are expected to anticipate the needs of their department and to place orders well in advance of actual need. They must also develop a system for the safe, clean, and orderly storage of such supplies and equipment. In most instances, they must use purchase orders and requisitions. Every technical training program should give some attention to the use of catalogs, order forms, and requisition procedures. Such knowledge can be helpful when a person is up for a promotion.

Much of the knowledge needed to hold a good job in the HVACR industry can be acquired in school, and the experience can be acquired through well-planned on-the-job training. However, no trade-technical school program will be all-inclusive. A great deal of the needed knowledge and information must be acquired by applying common sense to the routine tasks and problems that are encountered in day-to-day work activities.

Probably the most important step to be taken by people seeking careers in this field is to select the general branch of the industry in which they would like to specialize and then direct all their efforts to the accomplishment of these goals. The basic and technical educational requirements are about the same for all branches of refrigeration and air-conditioning sales, installations, operations, and services. To select a specialty does not mean that a person must stress one occupational field above all others, nor does it create a narrow and inflexible commitment to one field. Every technical training program should be subject to adjustment as need arises.

An ideal apprenticeship program closely correlates the class instruction with the apprentice's on-the-job daily work. A good plan for a company that does all facets of HVACR work is as follows (for a four-year program):

- 37 weeks—residential heating and cooling installations
- 37 weeks—residential servicing
- 21 weeks—commercial heating service
- 37 weeks—residential air-conditioning installations
- 37 weeks—residential air-conditioning servicing
- 21 weeks—commercial air-conditioning servicing
- 10 weeks—specialized work including the installation and servicing of humidifiers, electronic air cleaners, solid-state controls, chimney flues, and chimney dampers.

TECHNICAL EDUCATION REQUIREMENTS

The current opportunities for a formal apprenticeship in the fields of HVACR are rather limited. This circumstance has made it necessary for many people who want to qualify for a good job in this industry to turn to the next best thing: an informal apprenticeship. Under this "system," the student-worker must find a job and obtain whatever basic and technical education possible through his or her own efforts.

This industry has never stopped expanding since it was established, and its technology changes from year to year. However, many of these changes are concerned with applications of equipment; so the fundamental technical knowledge required stays much the same. For this reason, it is essential for the technician or mechanic to acquire a sound basic and technical education at the beginning of his or her career. With this background, it will be easy to adapt to new developments as they are introduced to the industry.

The chances for a formal apprenticeship are not good, and the chance of obtaining an on-the-job trainee position without some post-high-school technical courses is almost equally difficult. Therefore, the best entry into the field seems to be through a two-year course in refrigeration and/or air-conditioning technology, which is offered by either a public or a private trade-technical school.

Naturally, at the completion of a two-year college education, a person is not in the same position as the person who has completed a formal apprenticeship. One reason for this is that apprenticeship is generally four years and includes work experience. However, the person with the two-year associate degree might have higher lifetime earning potential and wider career opportunities due to more extensive education. However, he or she must first obtain the first few years of appropriate job experience. This can be comparable to a person with a four-year college degree in the professional areas.

A person who has an associate degree in the field is usually in demand by employers who hire engineering assistants, sales engineers, drafters, estimators, building engineers, plant engineers, field service managers, erectors, and many others. If none of these jobs sounds interesting, there is always the opportunity to establish an independent dealer-contractor business.

An additional advantage to having completed the two-year college-level education is that some people decide to continue their education. For example, a person who has earned an associate in science degree has demonstrated qualities that are admired by many employers. After a year or two of work, if a person decides to go back to school for a degree in refrigeration and air-conditioning engineering or mechanical engineering, the chances are good that the employer might subsidize part of the cost of evening classes on a part-time basis.

CHAPTER 3

WHERE TO TRAIN IN THE U.S.

There are various types of training available for a person interested in the HVACR careers described in this book. Each is explained separately in this chapter, which includes:

- apprenticeship training
- vocational high school classes
- evening classes
- on-the-job training
- correspondence courses
- advanced "skill improvement" programs

APPRENTICESHIP TRAINING

Apprenticeship training provides a very good training system for learning a skilled trade; it gives the trainee an opportunity to progress in an orderly manner from the entrance level to journeyman status through both classroom and on-the-job experience. Once the trainee is accepted into an occupation with an apprenticeship program, his or her progression is fairly well planned and assured—so long as he or she satisfies the requirements specified for advancement through the various stages.

Most apprenticeship programs cover a four-year period. The four years are usually divided into eight equal intervals, each six months long. Normally, the apprentice receives a set percentage of the journeyman's wage rate for each six months of training completed. The following is an example of an apprentice wage schedule typical of those found in collective bargaining agreements.

First six months—50% of journeyman's wage rate
Second six months—55% of journeyman's wage rate
Third six months—60% of journeyman's wage rate
Fourth six months—65% of journeyman's wage rate
Fifth six months—70% of journeyman's wage rate
Sixth six months—75% of journeyman's wage rate
Seventh six months—80% of journeyman's wage rate
Eighth six months—85% of journeyman's wage rate

Most apprenticeship programs also have guidelines for the on-the-job part in order to promote the growth of the apprentice's capabilities and at the same time to meet the ever-changing needs of our economy. Without this kind of organization, it is easy to overlook or neglect a training program's primary objectives. Some of the goals for the development of a well-rounded journeyman are pride in craftmanship, initiative, and ingenuity. These goals must be interwoven into the classroom work and the on-the-job part of the training program.

Apprenticeship is a training system based on a written agreement between the apprentice and the local joint apprenticeship training committee. Half of these members represent the employers and half represent the union. They are responsible for conducting and supervising the apprenticeship program at the local level. They test, select, and indenture the apprentice and register him or her with the U.S. Department of Labor's Bureau of Apprenticeship and Training or with the State Apprenticeship Agency. Local joint apprenticeship training committees often have training directors to assist them in carrying out these duties. Apprenticeship training committees are responsible for monitoring and evaluating the variety and the quality of each apprentice's performance. They are empowered to implement job rotation in order to ensure that the apprentice receives varied experiences with the latest materials, equipment, and construction processes.

Apprenticeship training programs have certain advantages over less formalized training programs. This is especially true of those programs that are national in scope, since they enhance the quality of the training through the use of standardized training materials. The completion of an apprenticeship gives the worker recognized status and dignity. It also provides a margin of job security, and often it may increase opportunities for promotion to foreman or another supervisory position.

VOCATIONAL HIGH SCHOOL CLASSES

Many high schools offer a variety of trade subjects, with some having vocational classes for juniors and seniors. These classes extend two to four class periods per day to give the student extensive training. In the larger cities, there are vocational high schools to which eighth graders may apply. Their primary emphasis is upon preparing young people for skilled jobs.

PUBLIC AND PRIVATE
POST-HIGH-SCHOOL EDUCATION

Local community colleges and area vocational schools offer one-year and two-year courses of study for various facets of HVACR work. Private schools offer similar programs, but they are usually only available in larger metropolitan areas. Sample courses of study are listed in Appendix A.

Some four-year colleges and universities offer associate and bachelor's degree programs in HVACR. Ferris State University at

Big Rapids, MI, generally has all graduates placed in jobs. Beginning salaries range from $25,000 to $36,000.

EVENING CLASSES

Many high schools and vocational schools offer evening courses that are planned to meet the needs of adults attempting to supplement their on-the-job experiences with related theoretical training. The students attending these programs have an adult point of view regarding the method of instruction and the contents of the curriculum. Attendance is voluntary and lasts only as long as the students think they are receiving the instruction they desire. The success of evening extension classes, therefore, depends mainly on giving students material relevant to their needs in industry. The methods of teaching and the range of materials covered should be as flexible as possible in order to meet the educational aims of the individual students.

The attendance of evening school classes can be a fruitful experience for students who are really serious about advancing in their chosen vocation. These classes help bridge the gap between the practical applications of the trade and the theoretical principles that, together, help form a better understanding of the mechanics and science involved. Students enrolled in these courses have the advantage of drawing upon the experience of the instructor. They can ask questions and participate in classroom discussions, which can be beneficial to the entire class. The instructor can, at times, shed light on a complex problem that may have arisen on one of the student's jobs and often can smooth the transition from a purely theoretical statement to one that incorporates base theory and practicality.

Usually evening programs are two years in length, although a few courses may cover a slightly longer period. The contents of these courses normally range from the fundamental aspects of the trade to an introduction to the more intricate theory and science associated with the field. These courses include laboratory or shop work.

ON-THE-JOB TRAINING

Another way to become a skilled tradesperson is by acquiring skills and knowledge through on-the-job experiences. On-the-job training (OJT) is usually an informal type of training, so it is usually not structured as well as the formal apprenticeship training program. The trainee learns the trade by performing a variety of tasks and by assisting other skilled workers in the performance of their duties. Quite often the trainee does not receive any additional off-the-job training to supplement the work experiences unless he or she purchases books to read and study.

This type of training naturally has drawbacks. It usually does not come under the auspices of a training committee, nor does the trainee enter into a written agreement covering a specified period of training. The lack of these two elements in a training program can present several problems that must be resolved if young workers are going to become well-rounded within the training period recommended for those particular occupations. If they do not receive the benefit of guidance from a policy-making body, their training can lack supervision and direction. Often journeymen are under constant pressure to meet production demands, and so they have little time to devote to the learner. And since on-the-job training is usually unregulated, the trainees do not receive the variety of work assignments essential for the development of competency. Therefore, the time required to advance from the initial entry-level job to journeyman status is usually greater than that required in formal training programs. In addition, many of the older skilled tradespeople are reluctant to help teach a young person—not wanting others to "know as much."

Unlike the apprenticeship training program, which provides for orderly progression and specified hourly wages for each level of competency attained, the wages paid to workers learning their trade through on-the-job training are normally determined by the individual's particular job classification and the type of duties performed. If administered in the proper manner, on-the-job training can lead to successful skill development. Adequate job rotation, monetary incentive for advancement, and job security must be included in the overall objectives of the training system if the training of skilled workers is to be accomplished effectively.

In some situations, on-the-job training offers workers who are too old to be apprentices the opportunity to acquire a trade. It also appeals to persons who are not interested in school work, since there is no related classroom instruction involved. Although this type of training has played a significant role in the training of skilled workers in some occupations, it has not been a very popular way of attaining journeyman status in the construction and related trades.

CORRESPONDENCE COURSES

Home study courses provide another way an individual may acquire knowledge and training to supplement on-the-job work experience. This type of independent study is widely known as correspondence education, and naturally it has both advantages and disadvantages. Perhaps one of its more attractive features is the freedom the student has in allocating spare time for study. However, this can present serious problems for students who lack initiative and determination. People who enroll in correspondence study courses must discipline themselves to use their spare time effectively. They should adhere to a study pattern that will allow them to accomplish their vocational objectives within the recommended and reasonable period of time.

ADVANCED PROGRAMS

Skill improvement programs usually are available for advanced training or training in a specialized area above the apprentice level, while the other training programs already described normally apply to training below the journeyman status. Journeyman training programs provide the worker with an excellent way of obtaining instruction on current trends and new techniques. They also afford the opportunity for greater knowledge and understanding of the applicable theory and practice associated with the skilled trade.

The technical knowledge gained through skill improvement courses often qualifies the participating workers for key and supervisory positions in their field. By concentrating on instruction in a particular phase of their occupation, journeymen can develop expertise in that area.

Skill improvement courses are usually offered by vocational-technical institutes. Quite often, these courses are sponsored by labor unions and equipment manufacturers, occasionally on a joint basis. Upon completion of the various units of instruction, the person is awarded a certificate that signifies satisfactory completion of the requirements for that particular course.

Manufacturers and distributors also hold classes and seminars for specific equipment, such as ice machine manufacturers, Amana, and Trane. Check with local distributors for details.

Contacts and new information can also be obtained from various trade shows such as the Design Engineering Show, the Plant Engineering Show, and the National Restaurant Association Annual Show.

LEARNING MATERIALS

You can request a book list that includes nearly every book and other publication available on the topics of air-conditioning, refrigeration, heating, and sheet metal from the following, which also publishes a fine monthly magazine for the industry:

Snips Magazine and Book Department
1949 Cornell Avenue
Melrose Park, IL 60160

CHAPTER 4

WHERE TO TRAIN IN CANADA

Although training programs vary in the different provinces, two sample programs for sheet metal apprenticeships are presented here, along with union and nonunion wages.

SHEET METAL TRAINING IN BRITISH COLUMBIA

Wages in 1995 are as follows:

- Nonunion trainee $11 an hour, approximately
- Nonunion skilled tradeperson up to $22 an hour, benefits vary
- Union wages

    1st year apprentice—50 percent of journeyman scale
    2nd year apprentice—60 percent of journeyman scale
    3rd year apprentice—70 percent of journeyman scale
    4th year apprentice—80 percent of journeyman scale
    journeyman scale—$26.53 an hour, plus benefits

For the union program, the pre-apprentice program provides twenty weeks of instruction. While an apprentice, there is an additional six weeks of instruction per year. Additional information is available from:

BCIT—British Columbia Institute of Technology
3700 Willingdon Avenue
Burnaby, BC Canada V5G 3H2

SHEET METAL TRAINING IN NEWFOUNDLAND

Nonunion trainee wages start at $4.75, which is the government minimum wage in 1995. Skilled-trades wages for nonunion work generally start at $15.00 an hour. This varies from company to company, depending upon the degree of skill needed for the specific job and the background and experience of the individual worker.

For union apprentices, the first ten months is the pre-employment training program. Then they go directly to the second-year apprentice wage. In 1995, these wages are:

Level	Commercial work	Industrial work
Second-year apprentice	$15.19	$15.53
Third-year apprentice	18.41	18.83
Fourth-year apprentice	20.54	21.02

During the second through fourth years, eight weeks of classroom work is provided per year.

SAMPLE PRE-APPRENTICE TRAINING PROGRAM

The following explanation is the program available through Cabot College for the sheet metal union training program.

This is an individualized ten-month certificate-level program designed to assist persons in developing sufficient basic skills and knowledge to enter the labor force as an apprenticed sheet metal worker. Following completion of the first phase of training, which leads to a certificate, the student may complete three additional periods of formal training (eight weeks each), preparatory to writing Interprovincial Journeyperson's Examinations.

Objectives

1. To develop an awareness of and concern for good safety practices in the workplace.
2. To develop skills and knowledge required for work as an apprentice sheet metal worker.
3. To develop and strengthen related knowledge and skill (technical and general) in subjects that complement and support the trade training.

Employment Opportunities

The graduate may obtain employment as an apprentice sheet metal worker with construction companies, service industries, and sales. Further training and experience may lead to employment as a supervisor, inspector, foreperson, estimator, and designer.

Entrance Requirements

Entrance requirements consist of a provincial high school graduation certificate or an adult basic education level III graduation certificate.

Program of Studies

No.	Course	Hours/week
XXX180	Trade Theory	8
XXX181	Shop Practice	12
XXX601	Communications	3
XXX621	Mathematics	3
XXX643	Science	3
XXX900	Welding	2
XXX956	Sheet Metal Drafting	3
	TOTAL	34

Course Descriptions

XXX180 Trade Theory The study of general shop practice; selection and specifications of materials used in the trade; capabilities of machinery and hand tools of the trade; basic principles of pattern development, parallel line, radial line, and triangulation methods; application of orthographic projection; basic metal working processes for locking seaming and edging, layout and fabrication, blueprint reading.

XXX181 Shop Practice Practical projects with emphasis on safe and proper work habits to develop the basic skills utilizing the theory outlined above.

XXX601 Communications This course is designed to teach trades students a range of oral and written communications skills. Through an emphasis on reading and writing fundamentals, students will enhance their workplace communication skills.

XXX621 Mathematics This course consists of the fundamental principles and concepts of mathematics needed by a sheet metal tradesperson who is required to solve and make measurements and layout. These mathematical principles and concepts are related to the general fields of arithmetic, algebra, plane geometry, and trigonometry.

XXX643 Science An introductory course in applied physics for the mechanical trades. Topics are selected from the appropriate field of study. Topics include: measurement (Si and English units), matter (physical and chemical properties), metallurgy, work, energy, power, force, motion, equilibrium, mechanics (simple machines), fluids, pressure, density, buoyancy, heat, temperature, expansion, gas laws, and electricity.

XXX900 Welding An introduction to the theory of spot welding oxygen-acetylene and arc welding, together with appropriate demonstrations and practice.

XXX956 Sheet Metal Drafting A course designed to provide instruction in the basic fundamentals of engineering drawing. Emphasis is placed on the development of drafting skills, geometric constructions, orthographic and auxiliary drawing, dimensioning standards, pictorial representations, sectional views, revolutions, developments, and intersections.

For further information, write to:

Cabot College
Sheet Metal Training Program
P.O. Box 1693
St. John's, Newfoundland
Canada   A1G 5P7

CHAPTER 5

PLANNING YOUR CAREER

Planning a career in the skilled trades involves careful preparation. The successes and failures experienced by young people are closely correlated to the way they plan for their vocations. Assuming that you are interested in the heating, ventilation, air-conditioning, and/or refrigeration (HVACR) field and that you have the potential, you should become acquainted with the various ways in which you can improve your chances of a successful career in these trades.

Selecting the right job and the best field of opportunity within HVACR is complicated due to the fact that this is a fragmented industry with many branches that are not closely related; each goes its own way and is concerned only with its own problems. The situation is further complicated because comprehensive counseling is not always available, and it is not possible to find someone who has adequate information that will give a clear picture of each field of opportunity.

Before any problem can be solved, it is necessary to define that problem and to discover the elements that make it a problem. In the case of HVACR the problem is to select the job that will offer the best chance to establish a rewarding career. To do this, it is necessary to examine the education, technical training, experience, and duties of many jobs—in other words, to outline the requirements and benefits of each field and each job.

Fortunately, you can refer to the many specific jobs described in this book. These jobs range from that of a semiskilled maintenance person who performs routine repair and maintenance, to that of a top-level supervisor who is responsible for all of the mechanical equipment in an entire complex of buildings.

The information on jobs represented here has been taken from actual interviews through questionnaires. For full information on civil service (government) jobs, the applicant should inquire at the specific agency. For information on jobs in a specific industry or company, apply at the personnel department of any business that employs people in this category.

JOBS IN MODERN CIVIC AND PRIVATE BUILDINGS

In the 1990s, the development of new civic centers, convention facilities, sports arenas, office buildings, hotels, and shopping centers has meant many fine job opportunities for HVACR mechanics and technicians, operating engineers, building engineers, and building superintendents. This is due to the fact that all these buildings have the latest machinery and systems for heating, ventilating, refrigeration, and air-conditioning. However, this growth has also brought many problems to hiring agencies and personnel managers who must recruit qualified people to service, maintain, and operate these complex mechanical systems, including heating, ventilation, refrigeration, and air-conditioning. This is primarily due to two facts:

1. The old buildings that are being replaced by these modern structures often had heating, ventilating, refrigeration, and air-conditioning plants that were rather primitive by modern standards. The people who were responsible for the maintenance and operation of these older plants did not, in many instances, have the technical training, experience, and skill

needed to manage the more complex mechanical systems. Employers often found it necessary to turn to other fields to recruit people with the needed qualifications.

2. Not enough people with the education and experience needed to handle the equipment at the operational level were available. As more and more large civic centers and private buildings were built, the competition for qualified workers increased. It soon became obvious that special training programs would have to be developed, but trade-technical schools were slow to respond to this need. However, this training void is now being filled, and some very good technical training programs are now available.

The four jobs outlined next do not cover the full range of jobs in this field, but they are typical and they point out the education, training, experience, and duties of this field. The best time to apply for these jobs would be while the buildings are under construction.

Chief Engineer. This is a supervisory position. The duties basically include planning and supervising the work of semiskilled and skilled workers in the operation, maintenance, and repair of heating, ventilating, refrigeration, air-conditioning, electrical, and other mechanical systems in a large building or group of buildings. It also includes being responsible for the care of stationary engines, boilers, compressors, pumps, cooling towers, evaporative coolers, condensers, steam lines, water lines, and other piping-system components. This position requires extensive experience in working with this equipment at lower level jobs, and past supervision of other workers.

Office Building Engineer. This is a technical and supervisory position. Its duties include operation, maintenance, and repair of low-pressure boilers and heating systems; refrigeration, ventilation, and air-conditioning equipment; and other mechanical systems under the direction of a chief engineer or other supervisor. The engineer also inspects and reports the condition of air, gas, water, and refrigerant piping throughout the building and makes emergency repairs and adjustments as needed. This type of position generally requires one year of experience performing the duties of a stationary firefighter in a building having equivalent mechanical equipment or two years of experience in the operation, maintenance, and repair of such equipment. But completion of a two-year college-level course in air-conditioning technology may be substituted for one year of the required experience.

Heating and Cooling Mechanic. This is considered a journeyman-level job. The duties include performing journeyman-level installation, maintenance, and repairs on heating, ventilating, refrigeration, and air-conditioning equipment and related components; diagnosing trouble and making repairs or replacements of mechanisms as required; performing minor and major overhaul of refrigeration and air-conditioning equipment and operational tests of equipment; preparing progress and operating reports; keeping a log or day book; and ordering spare parts and supplies as needed.

This type of position generally requires an applicant who has completed an apprenticeship in this trade or has had work experience equivalent to a completed apprenticeship. Mechanics should be able to perform duties of the job with minimal supervision. They also may be required to supervise the work of skilled and semiskilled workers in the cleaning, overhaul, and maintenance of refrigeration, air-conditioning, heating, and related equipment.

Auditorium or Sports Arena Engineer. This is considered a technical and supervisory position. It applies to plant operation and building maintenance in large buildings. The work involves maintenance supervision and operation of refrigeration and air-conditioning equipment; boilers; heating, ventilating, and electrical systems; and ice-making equipment (catering only). The engineer has overall supervision over building cleaning and building maintenance and assigns routine and special tasks to subordinates. The engineer recruits, trains, and supervises labor.

This type of position generally requires a job application, detailed résumé, interview, investigation, and examination. Quality of experience, where the applicant has shown the ability to assume responsibility, may be more significant than experience that called for routine activities. The completion of a two-year college-level course in refrigeration and air-conditioning technology is desirable.

SEMISKILLED JOBS

Only a limited number of formal apprenticeships are available in HVACR sales, installation, service, and operating fields. And not all of those wishing to enter this industry are able to complete a college-level course in refrigeration and air-conditioning technology. Fortunately, these are not the only means of entry into a good job in this industry.

In the past it was the policy of many employers, in both private and public buildings, to draw a rather clear-cut line between those who were responsible for the cleaning of buildings and grounds and those who were responsible for the maintenance, repair, and operation of mechanical equipment. In recent years, this line has not been so clearly drawn, and many employers now try to hire semiskilled workers who have the ability and the ambition to advance to more responsible jobs.

If an ambitious person understands the problem and elects to enter the building maintenance field through a semiskilled job, it is not necessary to remain on that level for many years. On the contrary, there is an ever-growing need for qualified workers to fill jobs in the new buildings, and employers are always looking for candidates who can be developed by on-the-job training and company-sponsored technical educational programs. In many instances the training programs use a combination of local trade-technical and home-study courses. The jobs described here show these types of opportunities in this field.

Building Maintenance Person. This is considered a semiskilled job. The duties consist of general custodial and light maintenance work in the upkeep of large buildings. Assignments might include repair or replacement of lighting-system components, doors, door locks, and similar tasks calling for handy-type skills; general cleaning of public areas; and setting up or removing facilities, desks, and cabinets. Duties may involve some leader activities when temporary employees are hired for special events in the auditorium and meeting rooms. The person who shows interest in learning can, with the guidance of a superior, work with a highly skilled person whenever possible and take specialized courses (evening school or correspondence) to develop an aptitude for air-conditioning, refrigeration, or another specific trade.

These jobs generally require a person who has completed high school and has some experience in maintenance and janitorial work. This person also must have the ability to carry out oral or written instructions, perform routine tasks without close supervision, use hand tools, and operate small tractors, fork lifts, and power-cleaning equipment.

Building Maintenance Foreman. This is a supervisory job, including supervisory work in the care and general maintenance of a large building or group of buildings. The foreman supervises, and sometimes participates in, the work of subordinates. All work is performed under the supervision of the building engineer or building superintendent. The maintenance foreman must have some knowledge of the operation of low-pressure boilers, heating, ventilating, refrigeration, and air-conditioning equipment and components; considerable knowledge of methods, practices, tools, and materials used in this type of work; and good knowledge of occupational hazards and safety precautions applicable to building cleaning and maintenance.

The experience requirements generally include graduation from a regular high school or vocational school and several years of experience in this field. This also includes the ability to plan, lay out work, and supervise subordinates in a manner conducive to full performance and good morale. Generally, a current employee is promoted from building maintenance person to foreman.

STATIONARY FIREFIGHTER AND OPERATING ENGINEER JOBS

Most of the newer buildings that have a large central mechanical plant have high-pressure boilers to provide steam for heating systems and hot water for domestic use. These boilers are in operation twenty-four hours a day, every day of the year. In most states, high-pressure boilers must be under the care of a licensed firefighter or engineer. Every building or complex of buildings has at least one firefighter or engineer on each shift. With one relief engineer, this means a minimum of four jobs in each plant of this type.

Stationary Engineer or Firefighter. This position is classified at the operator (shift work) level. Duties include supervising the operation of high-pressure boiler systems; operating, maintaining, and adjusting heating, lighting, refrigeration, air-conditioning, and ventilating equipment; inspecting, repairing, and reporting the condition of mechanical equipment throughout the building; keeping charts and records; and making written reports.

The requirements generally include one year of experience in assisting with the duties of a stationary firefighter or engineer in a large building or institution or equivalent experience in the operation and care of high-pressure boiler systems, heating systems, and related mechanical equipment. (Completion of a two-year college-level course in air-conditioning technology, which must have included the actual operation of high-pressure boilers and multizone air-conditioning systems, may be substituted for the required experience.)

ASSESSING YOUR OWN ABILITIES

How do you go about assessing your abilities? One suitable and effective method is to carefully analyze all the qualifications considered essential for admission and achievement in the field you choose. Also, you may ask yourself the following questions:

1. What is my real attitude toward this type of work?
2. Is it too mechanical, technical, or strenuous for me?
3. Is my interest in the occupation genuine or am I merely looking for employment?
4. Can I take orders and grasp instruction readily?
5. Will my temperament and personality allow me to work effectively as a member of a crew?
6. How do I feel about the related classroom training associated with the job?
7. Do I look at the training as an opportunity or as a requirement?

This is a sample of the questions that you, as an individual, can raise and attempt to answer honestly. When you have analyzed your answers, you should be able to determine whether you have a real desire to enter this field.

SEEKING THE SERVICES OF A
GUIDANCE COUNSELOR

As our nation progresses technologically, choosing an occupation becomes increasingly difficult. Therefore, it is often desirable to consult with a counselor about your choice. Faced with an array of job possibilities, you need to learn about those opportunities that will neither frustrate you nor waste valuable time and effort. Guidance counselors can help you recognize your personal characteristics and capacities and help you determine whether you are fitted for the particular type of work that interests you.

Based on conversations with you, your school record, results from testing, and other pertinent information, the counselor can help you arrive at a meaningful appraisal of your capacities, aptitudes, personal traits, and occupational inclinations. From this background of information, the counselor is able to recommend specific fields of employment in which you are likely to succeed. So long as a technical trade is plausible, you should continue your interest in the air-conditioning and refrigeration industry.

EDUCATION

In addition to its importance in competing for a job, education is a highly marketable commodity. In 1994, a high school graduate could look forward to earning $400,000 more during his or her working career than an individual having less than a high school diploma. In addition, a person with advanced technical training can expect to earn $200,000 to $400,000 over and above the earnings of a high school graduate (based upon a forty-year work life with fifty weeks of work per year). This gap will undoubtedly widen as employment growth continues.

General Education Development Test

In 1945 a program known as General Education Development (GED) was set up to accommodate returning veterans who had not completed their high school training. The American Council on

Education developed a GED test battery that covers the following five disciplines:

1. correctness and effectiveness of expression
2. interpretation of literary materials
3. general mathematical ability
4. interpretation of reading materials in social studies
5. interpretation of reading materials in natural sciences

Since its inception, the scope of the program has broadened to the extent that many states now allow anyone nineteen years or older to participate. Each state department of education establishes the policies and procedures for an adult resident to earn a high school equivalency diploma based on the results of the GED test. The state department of education determines the minimum test scores required to attain a diploma. It also establishes such criteria as age, residency, and any previous high school enrollment requirements for admission in the program. These regulations vary from state to state. If you are interested in this program, contact your state's Department of Education for specific details about its administration.

Home Study Programs

Another way to obtain a high school diploma is through correspondence course instruction. A number of reputable institutions provide home study educational programs, and enrollees can do quite well with the courses if they persevere. The student is, of course, primarily responsible for his or her rate of progress.

VETERANS' BENEFITS

If you are a veteran of the armed forces, you are entitled to education and training under the Veterans' Readjustment Benefits Act of 1966. This law provides the most important source of aid to veterans. Under it, they may receive benefits while training on the job or going to a high school, trade school, or college. To encourage dropouts to complete their schooling, the 1966 Act included a special provision permitting veterans to receive benefits while taking remedial courses leading to the completion of high school without affecting their entitlement to benefits for vocational training or college.

For servicepeople of the post-Korean conflict period and Vietnam era, different levels of benefits are provided for education and training. The amount of the benefit is determined by the following factors: the type of educational/training program; whether it is full-time, three-quarter-time, or half-time; and the number of dependents the veteran has. Under an apprenticeship program or in an on-the-job training situation, a veteran with no dependents would receive monthly benefits.

Cooperative educational/training programs for which GI benefits are provided combine formalized education with training in a business or industrial establishment, with emphasis on the institutional portion.

The Department of Veterans Affairs publishes a booklet, which it refers to as fact sheet IS-1. It contains much useful information concerning most federal benefits for veterans and their dependents and beneficiaries. Also, your area regional office of the Veterans Affairs Department can furnish you with additional details on other special services available to veterans.

EMPLOYMENT SERVICES

In addition to your educational planning, you will want to become familiar with several key sources of useful employment information and assistance. State public employment offices, which are affiliated with the U.S. Training and Employment Service of the U.S. Department of Labor, offer the potential job seeker service in four pertinent areas. They are:

1. job information
2. employment counseling
3. referral to job training
4. job placement

Trained personnel give applicants information about the various jobs available in their geographic area, specific job requirements, opportunities for advancement, rate of pay, and other related data. Employment counseling also assists both beginners and experienced workers who wish to change occupations. The primary purpose of this counseling service is to help job seekers become aware of their actual and potential abilities, their interests, and their personal characteristics.

LOCATION OF EMPLOYMENT

The location of employment for HVACR workers is characterized by the branch of the industry in which they are employed. For some sectors of the trade, the job site remains relatively constant, while in others the place of work changes and is governed by building and construction activity. Both the associated geography of an area and its natural resources have a bearing on the type of employment likely to be found in a given locality.

Highly industrialized areas provide substantial employment for maintenance, repair, and construction workers. Port cities (cities having navigable waterways and shipyard facilities) offer employment to servicepeople in marine refrigeration.

In summary, the type of industry, the available resources, and the natural features of the land influence the types of employment associated with a given locality. But jobs within the air-conditioning and refrigeration field are available in all city and rural areas—wherever there are people—and the volume of jobs available is closely correlated to the population of the area.

EMPLOYMENT SECURITY

The degree of employment security one has in the HVACR industry depends on several general and specific conditions. The state of our nation's economy, weather conditions, and other seasonal factors each have a bearing on the level and stability of employment in some specific jobs. The HVACR industry, like other industries, is affected by each of these elements; and some branches are influenced more than others by a particular condition or set of conditions. Moreover, while general economic conditions are likely to affect employment in all branches of the industry, some segments will experience less fluctuation than others; for example, most buildings need basically the same supervision and work on their heating and ventilation systems regardless of the state of the economy.

Employment in the construction industry is probably more affected by seasonal conditions than most of the other branches of the industry. The lack of building and construction work has been a continuous problem for the construction worker and has contributed to the underutilization of construction resources—especially human resources. To alleviate seasonal fluctuations, the federal government has adopted a counter-seasonal policy—a program designed to help strike a balance between the lack of construction activity during the winter season and the peak activity of the summer season. Fortunately, employment for the air-conditioning installers in construction is not quite as unstable as some of the other building trades, since a substantial part of their work is performed indoors.

Normally, those in maintenance and repair work on a steady basis. In other words, they seldom are required to take time off without pay. Generally, their work is unaffected by inclement weather conditions, and even during business lulls their employers are usually able to keep them gainfully employed in repairing or in servicing the equipment and apparatus they use in carrying out their job functions.

Seasonal changes have little influence on marine refrigeration employment; but work in this field is governed, to some extent, by the amount of shipbuilding taking place. Shipbuilding, like other construction work, lacks constancy. But since many marine refrigeration workers also seek employment in the repair and service specialty of the marine field, a large number of them enjoy reasonably stable employment.

Employment with a specific installation and/or service contractor varies with his or her specific line or lines of specialization and the weather conditions of the geographic area. As is obvious, heating equipment requires the majority of its service and repair during the months it is in use. The same applies to air-conditioning and ventilation equipment. But the company that handles both (as most do) can have a relatively steady work force with seasonal increases in overtime.

Job Openings

As the population rises, there is automatically a proportionate number of additional skilled people needed in HVACR work. Also, jobs become available due to each of the following:

- retirements
- occupational shifts
- upgrading (promotions) and specialization
- deaths
- other labor force separations

Thus, the total number of job openings for a particular occupation depends both on industry growth and worker replacements.

Hours of Work

Workdays and workweeks of HVACR workers mainly depend on the branch of the industry in which they are employed. Construction workers usually work straight days, while many workers in other branches of the industry work on a shift basis. Straight-day work generally means eight hours of work daily Monday through Friday, with a half-hour lunch period.

More and more labor agreements provide for work weeks of fewer than forty hours, and the number of such contracts continues to grow each year as more and more workers express a desire for longer periods of leisure time. Some are 37 1/2 hours (7 1/2 hours a day); some are 35 hours (7 hours a day); and even some are 32 hours (8 hours a day for only 4 days).

Employment and Earnings

Traditionally the earnings of workers in the HVACR industry are as good as or better than the earnings of workers in similar industries. The annual earnings of workers in these branches of the industry usually provide sufficient income to maintain a comfortable standard of living. No sample salaries are listed here due to the extreme increases in recent years, mostly just keeping up with our economy's inflation.

Most people in this industry work on an hourly wage rate. Wages and earnings depend upon several factors:

- the particular type of work being performed
- whether the worker is covered by a union contract
- the geographical section of the country where the work is being performed
- factors such as holiday pay

Fringe Benefits

A typical contract in the construction industry provides for substantial fringe benefits, and these benefits play an important part in contract negotiations. As a result, these workers are enjoying longer vacations, more paid holidays, better pension and insurance plans, additional sick and bereavement leave, and other benefits such as supplemental unemployment compensation and severance pay allowances.

The number of paid holidays continues to grow. A recent survey indicated that a large percentage of labor contracts now provide for nine or more paid holidays a year. The same study showed an increase in the number of workers receiving ten or more paid holidays per year. Also, many workers receive premium pay for holidays worked.

People working for small companies in heating, air-conditioning, and refrigeration installation and servicing generally receive fewer benefits, but the benefits they receive are still adequate.

OPPORTUNITIES FOR MINORITIES AND WOMEN

The opportunities for ethnic minorities in most facets of the air-conditioning and refrigeration industry are very good. For example, in recent years most labor unions have accepted into their memberships large percentages of minorities and some women to make up for the void over the past years.

Further specific information for individuals is available from the following associations:

Coalition of Labor Union Women
15 Union Square
New York, NY 10003

National Association of Minority Contractors
1333 F Street, NW
Suite 500
Washington, DC 20004

National Association of Women in Construction
327 S. Adams St.
Fort Worth, TX 76104

The U.S. Equal Employment Opportunity Commission indicates that "Minorities and women have improved their employment status over the years, but progress has been slow. Monitoring the performance of private employers is one of the Commission's means of ensuring that improvements continue to be made."

Additional information and reports are available from the Superintendent of Documents, U.S. Printing Office, Washington, DC 20402. You may request the office's free listing of materials on the topic of minorities, which includes the following publications:

Minorities and Women in Private Industry

Minorities and Women in Referral Units in Building Trade-Unions

OVERVIEW OF TOMORROW'S JOBS

Through the 1990s, the job structure for skilled workers in the HVACR industry has not changed drastically. Nonetheless, the contents of most major job classifications within the industry will experience some degree of change as new products, equipment, and machinery emerge. New installation methods and other technological accomplishments will also have an influence on industry employment and occupational requirements. Knowledge and skill requirements have been increasing as equipment for HVACR has changed significantly.

The increasing use of prefabricated components and modular construction techniques will probably gain impetus especially in the areas of pre-assembled industrial equipment and residential construction. The net result will be more hours of work performed in an in-plant environment and fewer at the job site.

Solving environmental problems, advancing space exploration and travel, alleviating transportation problems, and eliminating the uneconomical use of our natural resources will each have a major role in upgrading our standard of living and thus shaping our way of life. Activity in these areas will result in the creation of new jobs and in gradual changes in the structure of existing jobs.

In the next ten years, it will be necessary for our government to further develop comprehensive policies and programs that will enable our country to become self-sufficient in terms of our energy requirements. As oil becomes scarcer, the need for alternative sources of energy will become imperative. Also, it will undoubtedly become necessary for us to examine the present use we make of various sources of energy so that we may assign priorities to their utilization.

The continued fight for clean land and air and unpolluted water will require additional long-range planning and policy decisions by our country's leaders if we are to accomplish our objectives. The success of our endeavors will be influenced by decisions of people constructing both large and small buildings, which affects skilled workers in all trades and related fields.

Today, many of our major cities have outmoded and inadequate transportation systems. Consequently, millions of automobiles are used each day to transport people to and from work. This not only compounds our transportation problems, but also adds to our environmental problems, since these automobiles are responsible for the emission of thousands of tons of pollutants in the atmosphere daily. The antipollution progress made in the latter 1980s and early 1990s will continue. In addition, numerous cities will continue installing modern transportation facilities. The federal government has also become more active in this field; for example, the U.S. Department of Transportation is experimenting with an electrically propelled prototype vehicle capable of providing air-cushion rides at very high speeds. This vehicle is completely controlled by electronics through the use of computers.

Consequently, many new jobs will become available for those seeking employment in the technical trades. Also, space exploration and travel should continue to provide new employment opportunities.

In summary, the next ten years should see the opening of a considerable number of new employment opportunities in addition to the normal annual job openings that occur in the industry.

YOUR OWN ANALYSIS OF JOBS

Before making a career choice, be sure to consider the important alternatives and comparisons of several jobs or careers in which you are interested. An excellent means for making this comparison is to draw a chart like the one shown below and fill it in as completely as possible for each job considered. It includes the major considerations for any career.

Job Survey Chart

Job Title: _____

_____

Types of Companies: _____

_____

Locations of Companies: _____

_____

Job Description: _____

_____

_____

Education Required: _____

_____

Experience Required: _____

_____

Entry level salary $_____ per week.

Salary after two years experience $ _____ per week.

Hours per week of work: _____

_____

Fringe Benefits:_____

_____

Working Conditions:_____

_____

CHAPTER 6

EMPLOYMENT OPPORTUNITIES

People seeking jobs as apprentices, helpers, technicians, mechanics, and operating engineers in HVACR fall into three basic groups. The first group includes those who have completed some type of technical education and have had enough on-the-job experience to qualify as a mechanic, technician, or operating engineer. The second group includes those who desire a career in the refrigeration and air-conditioning industry but who have just completed their technical education and have limited or no related work experience. The third group includes those just entering the field seeking a job as an apprentice or trainee.

The employment opportunities of each of these groups are quite different, and this should be clearly understood. The first two groups who have completed a technical education and have had a few years of field experience will find many jobs open to them. Even more important, these people have a good understanding of the job opportunities in the field and know where the best specific jobs for them are to be found.

WHERE TO START

People who are just starting their technical training and have had limited or no work experience may find that jobs that will help them master one of the trades in this industry are not as easy to obtain. The old story about having to have experience in order to get a job, but not being able to get the experience without a job, is more than a story—it is a reality. However, it is a reality that has been mastered by many in the past, and it might be considered the first real test of a person's determination to succeed in the career of his or her choice. The information in this chapter is to help people trying to get a start in this line of work.

It is a fact that many employers try to hire people who are fully qualified with training and experience for the work they are to do, but not all employers follow this policy. Fortunately the many specialty branches of refrigeration and air-conditioning sales, installation, and service have jobs that call for different degrees of skill and technical education.

For example, extensive experience in a special field can be a handicap to technicians, who find it impossible to unlearn the work habits and procedures acquired in other jobs. For this reason, many employers are willing to hire a worker who has had limited training and experience if that worker has proven his or her intent to begin a course of formal study. Under these conditions, an employer can direct the educational activities to suit the needs of the specific job.

Another advantage is the fact that many jobs in the refrigeration and air-conditioning industry do not require extensive training or work experience. Therefore, many of these secondary jobs allow a student-worker to gain valuable experience.

WHAT TO EXPECT

The consideration of pay rates for the student-worker is important; you must adjust your initial expectations to the realities of the situation. This does not mean that you should allow an employer to take advantage of you by accepting very low wages in order to gain experience. However, do not lose sight of the most important fact: any job held during the training period should have one primary purpose—to provide on-the-job experience that will enable you to make practical use of the technical knowledge being acquired in your school work.

A few months of work in a busy shop, where you are allowed to work with mechanics and technicians on the actual installation, service, and operation of equipment, could have more career value than two or three years of work in a shop where all the jobs are routine and offer little chance to gain a variety of experience or to exercise your own initiative.

In many instances the beginning student in refrigeration and air-conditioning cannot offer an employer much more than two strong arms, a willingness to work, and an eagerness to learn. Fortunately, there is a place in many shops for people in this category. However, to advance in your career, you must plan to immediately progress from this classification. It is important for a new employee to show a good attitude, the ability to get along with fellow employees, interest in the business, initiative, and a measure of responsibility.

It is difficult to quote the wage rates a student-worker might expect to receive, but the wage patterns used in a formal apprenticeship can be used as a guide. In most instances, the apprentice is paid a starting wage that is fifty percent of the journeyman rate, with regular raises at six-month intervals until he or she graduates to journeyman rank.

LONG-RANGE GOALS

When planning a course of training and work experience that will take several years to complete and that will lead to a good lifetime career, you must make your decisions very carefully. If you decide upon HVACR, be sure to select the specialty within the industry that will offer the best opportunity along with being realistic for successful completion. For example, the education and training program should not be too rigid, and it should be within the limits of what is reasonable and possible for you. Also give some thought to the job and business opportunities that will be open in the geographic area where you plan to make your home.

Other chapters of this book indicate the education, training, and experience requirements for many jobs in HVACR sales, installation, service, and operation. Initially consider jobs in any or all of these fields. Then consider the job-hunting methods suggested next.

WHERE TO GO FOR THE FIRST JOB

Consider several of these job-hunting methods:

1. Personal visit to local HVACR shops. Remember that many entry-level jobs in the refrigeration and air-conditioning service field will be found in shops that employ only a few workers. You obtain these jobs by direct application to the employer. Although initial appearance and attitudes are of immediate significance, in the case of a student-worker the most compelling element in any job application might be the fact that you are trying to build a long-range future in the business, showing evidence of your enrollment and/or progress in a technical training program. If jobs are currently scarce, canvass every sales, installation, and service shop in the area. Then expand your effort to include every business and industry that employs mechanics, helpers, or operators for this field. This includes ice and cold-storage plants, meat-packing plants, food processing plants, transport refrigeration shops, and auto and truck-cab air-conditioning shops. Here again, enrollment in a trade technical training program could help secure the job.

2. HVACR supply stores. Supply stores that handle HVACR parts and components play an important part in the exchange of information that could lead to a job. The owners, managers, counterpeople, and clerks are in constant communication with the key people in every phase of the local industry. They know the employers who might want to hire a helper or mechanic. They know who might like to change jobs. And, they also know the qualified and experienced workers who are looking for a job.

3. Hotels and hospitals. Hotels and hospitals offer excellent on-the-job training to further develop your knowledge of various types of equipment. They use nearly everything you would need to work on in commercial refrigeration, air-conditioning, and heating systems.

4. Social and business contacts. Consider your family's list of friends, acquaintances, business associates, and relatives to determine if any are in a position to offer employment, give advice on the subject, or give a character or work recommendation that might be of some value.

5. Civil service jobs. Most city, county, state, and federal civil service commissions hire apprentices, helpers, technicians, and mechanics for both the refrigeration and air-conditioning trades and for closely related jobs such as maintenance workers and operating assistants. In addition, many of these commissions have an employment policy that includes part-time work for students, especially for the summer vacation months when extra help is needed. The fact that an applicant is enrolled in a trade-technical training program can show that the applicant is making progress, and that will enhance his or her prospects for employment.

6. School counselors and employment offices. A high school or college counselor can frequently offer advice due to contacts accumulated or experience gained from guiding other students.

7. State employment agencies. These employers can offer advice appropriate in your geographic area and suggest specific employers to contact.

8. Private employment agencies. Some companies (large and small) call upon an agency to locate the employees they need. When selecting a private agency, first compare several by asking who pays the fee and what their policies are. Also try to get a good report on their reliability from another person you know.

9. Veterans and armed service employment agencies. If an office is nearby and appropriate for your needs, the people there certainly can help.

10. Local chapter of a national trade association. For example, the Refrigeration Service Engineers Society has approximately 300 local chapters. It is a nonprofit educational organization, and its members hold key jobs in every branch of the refrigeration and air-conditioning sales, installation, service, and operation fields. Each chapter works very closely with the public and private trade-technical schools in the area. If there is a local chapter, application for student or associate membership may be made through the local secretary. For information write to Refrigeration Service Engineers Society, 2720 Des Plaines Avenue, Des Plaines, IL 60018.

RÉSUMÉS, INTERVIEWS, AND APPLICATION FORMS

Job application forms are important when seeking a job, especially when the applicant is dealing with employers who hire through a personnel department. There is no way to avoid filling out such forms, so the best approach is to make it as easy for yourself as possible. How do you do this?

It is a good idea to fill out a sample job application form before going on the first interview. You can do so by obtaining a sample form from local hiring agencies or from your local post office, where application forms for federal employment are available. All you have to do is ask for one. Other sources are city, county, and state civil service commissions and the personnel department of any business or industry.

You must be complete and accurate when filling out a job application form. Print or write as neatly as possible. When the person making a decision comes across an application that is hard to read and understand, he or she will not spend a lot of time trying to decipher the intended meaning. It is much easier just to put the application aside and reach for one that is clear.

When seriously seeking a job, give much thought to your attitude during an interview. It is best not to be too aggressive. Allow the interviewer to take the lead and ask the questions. Do not avoid the immediate question, and do not volunteer too much information. If you have submitted a résumé, the interviewer has most of the information he or she needs, and the aim now is to determine whether the résumé is accurate or exaggerated and to form an opinion of your personality.

A résumé has become more significant for experienced workers seeking the competitive higher level positions, especially on a supervisory level. If the résumé is poorly constructed, the applicant may never get as far as the interview. Many people seek professional aid in preparing a résumé, and books are available with sample résumés for different types of jobs. (Two good resources for putting together a résumé include the VGM books, *How To Write a Winning Résumé* and *Résumés for College Students and Recent Graduates*.) The primary purpose of the résumé is to provide more detail about your job experience, training, and education than is allowed on the application form.

One of the sources of higher level job information is the classified ad sections of industry trade journals and publications. To answer any of these ads in the classified section, you should generally send a letter and job résumé. It is not necessary to mail to only one employer at a time. The job résumé can be photocopied, but a separate covering letter should be typed for each employer.

Trade journals and trade magazines for the HVACR industry publish help-wanted ads; however, most of them are somewhat specialized and cover a limited branch of the industry. The one publication that covers several branches and also carries a large number of help-wanted ads is *Air Conditioning, Heating, and Refrigeration News.*

CHAPTER 7

AIR-CONDITIONING AND REFRIGERATION INSTALLATION AND SERVICING

Installation and servicing of air-conditioning and refrigeration equipment are very compatible, as they basically require a person with the same knowledge, experience, and skills. Being a good serviceperson is demanding, requiring the ability to think quickly and detect exactly what is not functioning properly. It is a challenging experience for the right person.

For the business owner, it is naturally helpful to build and maintain a good reputation in both installation and servicing, with capable employees both doing the physical work and dealing with the customers. Malfunctioning equipment is the best source for sales of new equipment; and prior installations are the best source for service calls—which points out the importance of employees' attitudes and the impressions they leave upon the customers.

AIR-CONDITIONING AND REFRIGERATION INSTALLATION

Air-conditioning and refrigeration *mechanics* install various types of systems from small air-conditioning units to walk-in coolers and frozen food units. This job requires mechanics to go to factories, supermarkets, hotels, stores, restaurants, and even construction sites. At a building site, a mechanic needs to follow blueprints and manufacturers' guidelines.

Each cooling or refrigeration system generally involves more than one machine, so mechanics need to understand the parts and how they fit together.

Air-conditioning and refrigeration *technicians* work with the equipment that controls temperature or humidity in an area. Some work for the companies that manufacture the equipment, helping design or test the equipment. Most work for companies that install and/or service the equipment.

AIR-CONDITIONING AND REFRIGERATION SERVICING

A large portion of residential and commercial air-conditioning and refrigeration work involves servicing current equipment that was previously installed by the same contractor or installed by someone else.

Being capable of handling service calls requires complete knowledge of installation of all the equipment, plus knowledge of all the other functioning parts not directly "worked on" during installation. A beginner can only do servicing after completing adequate courses, or by being permitted by the owner to work as an assistant with a qualified serviceperson. These are the best ways to become qualified as a serviceperson.

Service and Maintenance Contracts

Service and maintenance contracts for refrigeration and air-conditioning equipment have been in use for many years, but in the past most contracts covered the larger commercial and industrial installations. Owners and managers of these costly and complex systems understood the importance of having the skilled mechanics and service facilities of an established contractor readily available.

The same need for consistent service has transferred to residential units. Many installation contractors use their service department to build good will and increase profits. In fact, getting new service customers for equipment they did not install is a good long-range plan—to have them as satisfied customers when they do need new equipment.

With a large number of active service accounts, the contractor must have an efficient service organization. It is necessary to have a crew of well-qualified technicians and mechanics, service trucks, equipment, tools, shop, supplies, supervisors, and office force in proportion.

Service and maintenance contracts for refrigeration and air-conditioning equipment range in coverage from twice-yearly inspection to a broad insurance policy that covers every item or parts, equipment, and labor. The wide-range policies, however, are more common on the larger commercial and industrial installations and on out-of-town accounts.

The most common type of service contract covers refrigeration and air-conditioning equipment in homes, apartments, and small commercial establishments. Many contracts cover air-conditioning only, but in some installations the heating and cooling systems are combined and must be serviced as a complete unit. Some contracts cover equipment in bars, restaurants, and food stores where other types of refrigeration units are in use, and an effort should be made to include this equipment in the service contract.

Probably the best service contracts, from the standpoint of both the contractor-serviceperson and the customer, are the ones that call for inspection of equipment four times a year. The rate for such contracts varies from place to place, but one widely used method for figuring a base rate is one hour of labor per call. Thus, if the charge rate is $55 per hour, and the contract stipulates four calls a year, the annual charge is $220.

The annual charge for contract service includes travel time and mileage to and from the job (except under unusual conditions). The contract covers inspection, adjustment, cleaning of the condenser and unit, oiling, and minor repairs, but it does not cover the cost of filter pads unless so stipulated. In case of a serious breakdown of the equipment under contract, when major components or controls have to be replaced, the usual rate for labor is charged plus the normal markup on parts, components, and supplies.

The established dealer-contractor who sells and installs equipment seldom has time to actively solicit additional service customers during the busy season, so this is a good time for a new independent contractor-serviceperson to enter the field. There are many ways to acquire service customers, and the new contractor should use all of them. In the first days of independent operation, however, the best chance is with older units and with overflow work from the large dealer-contractors, chain stores, and out-of-town sales agencies.

Strangely enough, the chief complaint of disgruntled customers is not about the cost of service. Few customers balk at a reasonable charge, especially if they received good and prompt service. The one thing that attracts and holds a customer is the assurance of being given prompt attention when he or she calls for service—especially in an emergency.

SUBURBAN AND RURAL SERVICE

The contractor-serviceperson in the suburban and rural field is required to work on a wider variety of equipment, and many customers may be in the residential class.

The suburban contractor needs the same type of service truck, equipment, tools, and supplies, but should add pipe-cutting and pipe-threading equipment for pipe up to two inches in diameter. With these tools he or she will be able to do small plumbing and pipe-fitting jobs that are required on many refrigeration and air-conditioning installations. (In a city, such work might be subcontracted out.)

The opportunities for the success of a suburban refrigeration and air-conditioning shop arise by reason of the buying habits of millions of customers. Rural and suburban families tend to buy day-to-day household needs from local stores but travel to the city to purchase major items such as appliances, refrigeration units, air-conditioning, and heating equipment. All of this equipment will need service, and this need provides the opportunity for success for a qualified contractor-serviceperson.

This is not to imply that city-based dealer-contractors neglect their outlying customers. On the contrary, many such dealers sell quality merchandise and give first-class service, but they face certain built-in problems that make good service very hard to carry out, especially during the rush season. They must also charge higher rates due to the travel times.

Many families, both urban and rural, are addicted to bargain hunting, and they often buy refrigeration and air-conditioning equipment from outlets that seek high volume and quick profits and have little interest in quality installation procedures and service. Many volume stores offer a minimum of service while the equipment is under warranty but abandon all pretense of service when the contract ends. Of course, this is all good for the local service-oriented contractor.

Some chain stores maintain a fairly efficient service organization, but service is usually on a route schedule and response to an emergency call may be slow. In some instances the more reliable chains try to offset this deficiency by subletting certain service jobs, often the more difficult ones, to an independent local contractor-serviceperson. This arrangement can give a measure of security to a newly established business, but there are risks. The person who takes this route may learn, at the end of the first busy season, that hard work has benefited the chain store rather than the serviceperson.

All of these things can work to the advantage of the new contractor-serviceperson, but they can also work against him or her if the underlying factors that can contribute to success are not understood. The following outlines sum up the basic considerations.

Cost of service work. The single most important element that works to the advantage of the new contractor-serviceperson is the fact that he or she is able to operate and make a good profit on an hourly service rate that is well below the rate that must be charged by the large dealer-contractors. For example, a medium-sized dealer who employs five full-time mechanics and operates five fully equipped service trucks, also needs shop space, tools, equipment, parts and supplies, supervisory personnel, and office force in proportion. The hourly charge to the customer for service must be high enough to cover overhead and allow for a reasonable profit. The independent one-person operation does not have to charge for these overhead items.

Charge rates for service. Hourly rates for service on refrigeration, air-conditioning, and heating equipment vary from city to city. When such work is done after normal business hours the rate goes up from 50 to 100 percent. The following formula for setting rates is widely used: Take the base rate of pay for a journeyman, add fringe benefits, add 10 percent for lost time, and multiply by two. Thus, if the base rate is $25.00 per hour, add $4.00 per hour for fringe benefits to make $29.00 per hour; add $2.90 for lost time to make $31.90 per hour; then multiply by two to get a charge rate of $63.80 per hour.

Priority for service calls. Customer priority for service calls is one of the toughest problems faced by the large and the small dealer-contractor. It is never the intent of service contractors to slight customers, but they must, however, consider locations and driving time in scheduling their sequence of calls each day.

Service territory. Most dealer-contractors assign a mechanic and a truck to a given territory. Usually the territory extends no more than ten miles from the shop; and by working a district at a time and concentrating calls, a mechanic should cover from twelve to twenty calls a day. If the mechanic encounters major service problems, however, the number of calls can be greatly reduced. To avoid disruption of service schedules, many shops assign major service to a separate department, since it is so much more variable in time required.

Emergency service. When a dealer-contractor serves both urban and rural customers, problems can arise when rural customers demand emergency service during the peak season. If the customer is an old and valued one, the dealer must respond as quickly as possible. If the customer lives as far as forty miles from the shop, that means a minimum of two hours' travel time for the round trip. If two hours are needed to make the repairs, the mechanic will be away from the regular territory for four hours. If several rural calls are received in the course of a day or two—and they often are—a mechanic might become days behind schedule. When this occurs, and regular customers are without air-conditioning in the midst of a hot spell, these customers tend to become unhappy, but they would be even more unhappy if the dealer put the mechanics on overtime and doubled the rates. To complicate this situation, the efficiency of overworked mechanics may drop to a low level. The life of a dealer-contractor is not an easy one, but it can often be profitable.

The independent contractor-serviceperson. The contractor-serviceperson who sells technical knowledge and experience directly to the customer in the form of labor has lower overhead costs and should be able to do fine charging a rate well below that of the large competitor. If rate charges are based on a figure one-third below the competition, the serviceperson would arrive at a very advantageous hourly rate. Rural customers would not be charged extra for mileage (just the travel time), all of which would put the contractor-serviceperson in a good competitive position.

Sales. The independent contractor-serviceperson does not have a chance to see much new equipment or make many new installations, but he or she is able to supplement income with the profit from the sale of parts, belts, supplies, filter pads, and refrigerants.

These jobs are available throughout the country; homes, businesses, commercial and industrial buildings all need service on their air-conditioning, heating, and refrigeration equipment.

Earnings. Employment through the 1990s will continue to be good as the population and economy expand. Earnings depend on the type of equipment involved. Mechanics who install large units earn more than those who work with small home units. Under union contracts, the average pay is $20,300 a year, although some experienced mechanics earn at least $35,000 a year. Work is relatively steady, but there can be overtime during peak periods of summer heat.

CHAPTER 8

MAINTENANCE AND OPERATIONS

Maintenance is another field of opportunity in the refrigeration and air-conditioning industry that offers many fine jobs for qualified people. These jobs include the technicians, mechanics, servicepeople, operating engineers, foremen, and building managers who are responsible for the service and operation of mechanical equipment and systems used in modern office buildings, hotels, hospitals, schools, and other commercial and industrial buildings.

The education, technical training, and experience requirements for these jobs are similar to those already explained in earlier chapters. The following explains the job of a maintenance and operations foreman.

This is a very good job, and the route toward obtaining it is important to anyone who seeks a career in refrigeration and air-conditioning. Many of the people who hold these responsible jobs have a background of training and experience in refrigeration and air-conditioning sales, installation, operation, and mechanical service. Remember that many of the people who hold these jobs are of some maturity, and this does not necessarily mean a person of advanced years. It can mean people in their late twenties, or older.

For example, a typical position of a maintenance and operations foreman in a new government building is held by a person who has worked several years for a refrigeration and air-conditioning contractor servicing and maintaining a fleet of trucks and doing general building maintenance. During this time, the candidate has studied refrigeration theory and practice through a student membership in the Refrigeration Service Engineers Society. The potential foreman would be enrolled in the night school and home study classes it sponsors. After completing the three full years of work and passing the certificate examination, the journeyman level of training would be achieved.

However, after all this training, advancement to a better job with the present employer might not be possible; there simply would be no need for this knowledge and training. At this point, candidates could look for a job that would provide an opportunity to use the training and experience to better advantage, or they could attempt going into business for themselves.

The interested engineer would take a civil service examination for the job of general mechanic with the General Services Administration of the federal government. This examination includes many questions on refrigeration and air-conditioning, along with other mechanical subjects. Receiving a passing grade qualifies the candidate for placement on the civil service "list"; but it may take a four-month wait for an interview.

During the interview, the candidate learns of openings for refrigeration and air-conditioning mechanics and can ask to take an examination for one of those jobs. After a successful oral exam and interview, a position as a general mechanic is possible. The next step upgrades the employee to the position of maintenance and operations foreman, where full responsibility for the building is assumed.

When starting this type of job, a person must first learn the layout of the building and the mechanical plant. It is necessary to know the location of every office, conference room, storeroom, closet, stairway, and fire escape. For the mechanical systems and equipment, it is essential to know the location of every component, fixed or portable; every electrical switch and service panel; and every service shut-off valve. Whenever possible, a person learns this for two to three weeks under the supervision of the person leaving the position.

Are the mechanical systems in a building of this size extremely complex? A typical system for an office building includes a 62-horsepower, low-pressure boiler for the heating system and domestic water; a 175-ton refrigerated air-conditioning system; a cooling tower on the roof; an assortment of pumps for water circulation; several self-contained water coolers; and domestic refrigerators and other special equipment. The boiler, heating system, hot-water system, and air-conditioning system all have automatic controls and require little attention under normal operating conditions. In addition, all mechanical systems are equipped with the required safety devices.

Typically a maintenance and operations foreman has two to eight custodial employees and one to three general mechanics. The custodial employees do the cleaning and general upkeep work, and the general mechanics attend to all mechanical problems including minor electrical troubles, light-bulb changes, and plumbing calls. The foreman also supervises all mechanical work let out to subcontractors and helps out with the mechanical service if needed. Specialized work such as elevator service is taken care of by an outside service company, and some major overhaul and repair work is let out on bids.

Within each normal workday, calls are classified as routine, emergency, disaster, and catastrophe. In addition, some calls for service fall into the nuisance class. However, all calls must be analyzed carefully—some callers exaggerate!

Examples of nuisance calls are: something doesn't work because the electric plug was pulled out and not noticed; or a person is too hot or too cold. You must learn to handle these calls with courtesy, firmness, and psychology. Many workers think that the temperature in the building is controlled separately for each individual room. A building is usually divided into zones, and the temperature is controlled by zones. There is little we can do to change the temperature in a specific room. You must check the thermostat for that zone for malfunctioning equipment. If the temperature in the room is close to the allowable range, there is no problem other than one individual's desires. You can possibly change slightly the direction of air from an outlet.

In addition the job does require paperwork. This includes employee records; evaluations of employee performance; regular reports on every component of the mechanical and electrical systems; reports on water-treatment results; recommendations for preventive maintenance and repairs; obtaining bids for supplies and mechanical repairs; ordering supplies, parts, and tools; and arranging for safe and orderly storage.

In summary, this job certainly is a challenge and there are many problems, but it is rewarding with a good feeling of accomplishment.

The next job advancement would either be to a larger building or several buildings in a complex. This type of work has room for growth and advancement, especially for young people since every new office building, hotel, sports arena, shopping center, hospital, or school must have employees to hold jobs equivalent to this one. There is always a need for qualified technicians, mechanics, operating engineers, maintenance foremen, and building managers on every level.

Personality is an important factor in this type of job. If a person is a loner or has trouble getting along with people, he or she should seek a job where such ability is not required. That does not mean the person is barred from mechanical maintenance jobs in buildings; many jobs in this field do not require close contact with tenants or the general public. The attributes needed include a clean and presentable appearance, businesslike behavior, willingness to give much attention to details, and the ability to deal courteously with tenants, fellow employees, and visitors to the building.

CHAPTER 9

MARINE REFRIGERATION

Marine refrigeration and air-conditioning is a fascinating branch of the industry available only to people who live in an area where there are shipbuilding and ship repair facilities.

The United States does not have a large merchant marine, and a portion of its merchant ships are built in foreign yards; but the U.S. Navy, Coast Guard, and Army do operate every type of water transport vehicle ranging from nuclear-powered aircraft carriers to motorized scows. Each ship or boat that carries perishable cargo or has a crew living aboard requires mechanical refrigeration to preserve food and other perishables and air-conditioning for the office and living areas.

This field offers many jobs for helpers, apprentices, journeymen, and supervisors. Most supervisory jobs are filled from the journeymen level by means of promotional examinations in U.S. government facilities and on the basis of merit and seniority in private facilities. However, the education, technical training, and experience requirements are similar in government jobs and for jobs in privately owned shipyards, boatyards, and repair facilities.

A journeyman refrigeration mechanic works from blueprints and verbal direction to install, maintain, and repair refrigeration and air-conditioning equipment and components used for living quarters, public areas, and industrial cold storage. To become a journeyman, you must first complete a four-year apprenticeship in the trade or have had practical experience equivalent to an apprenticeship. This experience must include:

- at least six months in the operation and service of modern multizone air-conditioning equipment
- the layout, fabrication, and installation of refrigeration piping and tubing
- the use of freon refrigerants
- the installation and repair of compressors
- the installation of electric motors, fans, and blowers
- possibly pipe bending, tube bending, silver brazing (soldering), hand cutting, and threading of pipe

Well-rounded experience in industrial pipefitting may be substituted, on a month-for-month basis, for a maximum of half the required experience.

These qualifications for jobs in refrigeration and air-conditioning in a naval shipyard or repair facility also apply to jobs in private industry. There are four main classes of shipbuilding and ship repair work for both private and governmental yards. These are:

- building new ships and boats
- converting existing ships and boats to a different class of service
- major overhauling of ships and boats in service
- making voyage repairs usually of an emergency nature, done while a ship is in port for discharge and/or loading of cargo and passengers

REFRIGERATED CONTAINERS

Using containers to ship perishable products is a relatively new development in marine refrigeration, but it has grown at a very rapid rate. The shipping container is an insulated box that can be loaded at a distant point and hauled to the dock by air, truck, or rail. A widely used size is approximately eight feet wide by eight feet high by thirty-five feet long. A refrigerated container uses a mechanical refrigeration unit similar to that used for a truck or trailer body. Nonmechanical refrigerating systems are also used.

Before World War II, the fishing boats and tuna clippers that operated offshore were limited in the distance they could go due to the maximum amount of ice they could load in the hold. If the ice melted before fish were found, the boats had to return to port with empty tanks. Mechanical refrigeration greatly increased the range of these boats, and they could stay at sea for weeks if necessary. Boats of this type operate from any port where offshore fishing fleets operate. Servicepeople are naturally needed to maintain and repair the refrigeration equipment.

TRAINING FOR MARINE REFRIGERATION JOBS

One good way to prepare for a journeyman-level job in marine refrigeration is through a four-year or five-year formal apprenticeship; but there are only limited openings yearly. In addition, work experience gained in other branches of refrigeration, air-conditioning, and related fields is acceptable to many employers. This makes it possible to acquire the needed training and experience through trade or technical school courses and on-the-job experience. Supplement the standard refrigeration and air-conditioning technology subjects with special added emphasis on the following courses or topics:

1. *Refrigeration servicing procedures.* Variations from normal operating conditions are observed in the laboratory, and remedial service procedures are performed. Check-test-and-start practices for systems above five-ton capacity are practiced.

2. *Refrigeration and air-conditioning.* This course is offered for journeymen and technicians who desire additional technical information concerning refrigeration systems. Instruction is given in refrigerant properties, refrigeration cycle controls, evaporators, condensers, pumps, compressors, piping, and fitting.

3. *Refrigeration and air-conditioning controls.* This course is open to refrigeration mechanics and technicians only, and its purpose is to help these workers understand the various types of controls for refrigeration and air-conditioning systems. The course includes fundamental principles of electric circuits and pneumatic circuits, installation problems, and service and maintenance problems.

4. *Refrigeration welding and silver brazing.* This course is offered to supplement the mechanic's skill in silver brazing and soldering. It covers technical terms, torches, fluxes, metal alloys, brazing rules, practice in making brazed joints, and testing of completed work.

CHAPTER 10

COLD STORAGE AND INSTITUTIONAL REFRIGERATION

Cold storage and institutional refrigeration is not as widely known as other segments of this industry because it is smaller. However, it is an expanding field and offers fine job opportunities for people with the right education, technical training, and experience. Job classifications range from refrigeration plant operator to plant superintendent.

A good portion of the jobs in cold storage and institutional refrigeration are offered by city, county, state, and federal civil service commissions; and they are concerned with the management, operation, maintenance, and service of refrigeration and air-conditioning plants in public institutions such as hospitals, schools, and penal institutions. Comparable jobs are available in every business and industry that requires cold storage, food processing, meat packing, and other on-farm and off-farm activities that require refrigeration. The largest and fastest-growing segment of this industry is in the storage and distribution terminals that are operated in conjunction with transportation. This includes air, rail, truck, and marine cold-storage terminals and distribution facilities.

This chapter deals with several of the typical jobs available in this field. It does not cover the full range of jobs, but the examples are typical and describe the education, training, and experience requirements generally applicable in the entire field of employment in general.

REFRIGERATION ENGINEMAN

A refrigeration engineman is considered an operator and works on a shift basis, since these duties must generally be carried out twenty-four hours a day, seven days a week.

The refrigeration engineman works under direction, operating and caring for mechanical equipment used for refrigeration and cold storage of food and other goods. This work includes operation of refrigerating compressors, brine pumps, circulating water pumps, evaporative condensers, blowers, fans, control valves, and other components of large mechanical refrigeration systems. The refrigeration engineman makes repairs as needed; checks and regulates temperature, humidity, and forced-air circulation to meet the specific requirements of the goods in storage; records temperatures, pressures, and humidity readings in logs; and keeps a daily record of plant activities.

This position generally requires a person who has completed the twelfth grade and has three years of experience in the operation of mechanical refrigeration systems of not less than 100 tons of refrigerating capacity per twenty-four hours. The engineman must be able to read and work from plans, drawings, and mechanical or construction specifications and may be required to supervise the work of skilled and semiskilled workers in plant maintenance and overhaul. This job requires the ability to think and act quickly in emergencies.

REFRIGERATION MECHANIC

The refrigeration mechanic job is generally a union journeyman position (not apprentice). The person installs refrigeration components such as compressors, evaporators, condensers, motors, pumps, fans, and blowers. The mechanic installs and connects piping, fittings, and tubing for refrigeration and air-conditioning equipment; calibrates and repairs refrigeration plant controls; diagnoses operating malfunctions and makes needed repairs and adjustments; and tests, charges, repairs, and adjusts hermetic-type refrigeration units, self-contained refrigeration units, ammonia ice and refrigeration plants, and brine refrigeration systems. The refrigeration mechanic also frequently supervises the work of helpers and apprentices.

The position generally requires the completion of a recognized refrigeration apprentice training program of at least four years duration. If not union, it can require six years of experience in the installation, alteration, maintenance, and repair of commercial, industrial, and domestic refrigeration and air-conditioning units and systems, plus one year of experience on the journeyman level including experience in the installation, operation, and maintenance of large multizone air-conditioning systems.

Most two-level supervisory jobs in this field are filled by the journeyman level or by an experienced person who previously completed a two-year college course in refrigeration technology.

PLANT ENGINEER

A plant engineer works in a supervisory capacity. The person plans, assigns, and supervises the work of skilled, semiskilled, and inmate helpers (institution only) in the operation, maintenance, and repair of heating, ventilating, air-conditioning, refrigeration, and cold-storage equipment, including stationary engines, boilers, compressors, pumps, and condensers; water, steam, gas lines, and piping; and fire protection and safety equipment. Plant engineers evaluate staff performance and take or recommend appropriate action. They prepare requisitions for fuel equipment, spare parts, and supplies and are responsible for clean and orderly storage of such items. They keep or direct the keeping of charts and records and prepare written reports as needed.

The position generally requires a person with a minimum of two full years of recent experience in the maintenance and operation of large modern heating, ventilating, air-conditioning, refrigeration, or cold-storage systems of the type found in large commercial, industrial, or governmental institutions and buildings. The plant engineer must have a general knowledge of state safety orders and industrial safety regulations that apply to the operation of extensive mechanical systems, and he or she must be familiar with the principles of supervision in order to direct people effectively.

CHAPTER 11

AUTOMOTIVE AIR-CONDITIONING AND TRANSPORT REFRIGERATION

Modern automotive air-conditioning started shortly after World War II when enterprising individuals built mechanical refrigeration units from conventional components and mounted them in the trunk compartments of passenger cars. These units filled nearly the entire trunk area, and the air-circulation system required extensive penetrations through the inner body of the car. The controls were rather primitive, but the system did work.

As demand developed, engineers designed units that could be installed in almost any make or model of car, and production mounted. At the present time, automotive air-conditioning is, in terms of the number of new units sold each year, a fast-growing branch of the refrigeration industry.

Automotive air-conditioning is considered an accessory; therefore the sales, installation, and service of this equipment are more closely related to the automobile industry than to refrigeration and air-conditioning. Most of the mechanics who specialize in this field started out as automotive servicepeople and mechanics.

Automotive air-conditioning equipment manufacturers are now combining the heating and cooling system into a single unit with a more complex control system. These installations are more difficult to service and maintain; most are installed when the car is built, not added on later, as were the former air conditioners most popular in the 1950s. Therefore semiskilled installers are no longer needed in the industry. Manufacturer training programs have been extended to produce the qualified technicians who are needed to handle these complex systems.

Automotive air-conditioning is no longer a separate job as it is combined with the heating system of the vehicle (except for servicing the coolant, which is generally handled at service stations equipped to do this).

The best opportunity for a young person who wants to work in an automotive air-conditioning shop would be to attend a training school sponsored by an equipment manufacturer if convenient geographically. The schools are open to the employees of any shop where a substantial amount of equipment is sold, installed, and serviced. The classes and shop demonstrations usually last for three to five days, and there is no tuition fee. However, many employers are reluctant to send employees to these schools because they are required to pay wages and travel expenses. An ambitious person might find it best to ask a potential employer to sponsor him or her without paying any expenses.

TRUCK-CAB AIR-CONDITIONING

Truck-cab air-conditioning is another field that has grown in recent years, and the job opportunities are very similar to those found in automotive air-conditioning. Truck-cab air-conditioning is divided into two classes:

Class 1 includes pickup trucks, delivery vans, and light-to-medium weight commercial vehicles.

Class 2 includes all heavy-duty trucks.

The air-conditioning equipment used on large trucks is quite different from that used on passenger cars and light commercial vehicles, and the service and installation problems are more complex. For example, the compressor for a truck air-conditioning unit is usually mounted on the side of the main drive engine and gets its power from a pulley on the fan hub or auxiliary drive. The condensing unit is mounted on the top of the truck cab, with the cooling coil extending into the cab. This is a long run for the suction and discharge hoses and for the electric wiring. In addition, the road action of the truck body and vibration from the diesel engine can cause both installation and operation problems.

Many truck-cab air-conditioning agencies operate with a transport refrigeration shop, so the mechanics must be qualified in both classes of work for installation and service. Transport refrigeration shops are usually located near the off-ramps of main highways or expressways. Some are factory branch stops, but most are owned by independent operators. Obtaining a job in this field is similar to a job in automotive air-conditioning.

TRUCK TRANSPORT REFRIGERATION

Truck transport refrigeration is a highly specialized branch of the industry. In some respects, it has the same problems and possibilities as automotive air-conditioning. However, there is one important difference—truck transport refrigeration is much less subject to seasonal differences. The decision to have air-conditioning installed or serviced in a private automobile or commercial vehicle is made at the discretion of the owner, but transport refrigeration equipment is installed and maintained as a matter of necessity.

Automotive and Class 1 truck air-conditioning requires only limited skill and experience compared to the mechanics who install, service, and maintain transport refrigeration equipment. A large refrigerated truck and trailer unit may represent an investment of $30,000 or more, and the value of the cargo could easily double this figure. Truck owners are reluctant to entrust the care of such equipment to mechanics who are not fully trained and qualified.

To gain experience in this field (Class 2), a person usually begins by learning and working on Class 1 equipment, which is especially convenient if the transport refrigeration shop also does Class 1 work and Class 2 truck-cab air-conditioning, as many do. The busy season for these specialties coincides with the busiest season for transport refrigeration, since transport refrigeration equipment gets its hardest use in hot, humid weather. Therefore many shops have to hire extra help for the summer months. Jobs in this field are not limited to shops that accommodate freeway traffic. Many good jobs are available with the meat-packing and distributing companies, food processors and distributors, dairy companies and creameries, and with the big fleet operators such as railroad, truck, and shipping lines.

Many of the manufacturers of automotive air-conditioning, truck-cab air-conditioning, and transport refrigeration equipment operate schools to train people in this field of work. For information, you can write to any of the manufacturers of this equipment or apply at local installation and service shops. The following is a list of some of the better-known training sources:

1. York Corp., Division of Borg-Warner Corp., Grantley Road, York, PA 17405
2. Scott-Engineering Sciences, Co., 1400 Southwest Eighth Street, Pompano Beach, FL 33060
3. Thermo-King Corp., West Nineteenth Street, Minneapolis, MN 55420
4. Frigiking Co., Division of Cummins Engine Co., 10858 Harry Hines Boulevard, Dallas, TX 75220

NONMECHANICAL REFRIGERATION

There are several nonmechanical methods for refrigerating cargo transported by air, rail, truck, or ship. Many old-time refrigeration installation and servicepeople have a tendency to avoid or ignore all nonmechanical methods. However, with the development of new techniques for the use of liquid nitrogen in the transportation of perishable products, this method can become more widespread in its use.

The four principal forms of nonmechanical refrigeration are:

1. liquid nitrogen for both long-haul and local delivery
2. liquid nitrogen for food processing
3. holdover plate-type evaporators and a central refrigeration plant
4. nonmechanical refrigeration by means of dry ice bunkers and blowers

For information on liquid nitrogen transport refrigeration and training and employment information write to: Union Carbide Corp., Linde Division, 270 Park Avenue, New York, NY 10017.

CHAPTER 12

SALES CAREERS

Many people have the feeling that a *good* salesperson can sell anything. Although this is basically true, a knowledge of the specific products and/or services available is essential in this career. Therefore, the person selling air-conditioning and refrigeration needs to know the products, their functions, and operations; must have the technical knowledge to demonstrate the product; and must have had substantial experience with the people who will buy, install, and service the product. This is why manufacturers, distributors, and wholesalers in the refrigeration and air-conditioning industry often look for people with adequate experience such as mechanics and technicians when they need a salesperson. This is also why so many successful salespeople and sales engineers formerly held jobs as technicians, mechanics, estimators, and specialists in refrigeration and air-conditioning installation, operation, maintenance, and service.

It is difficult to even estimate the number of salespeople who are employed (directly and indirectly) in the sale of parts, supplies, chemicals, refrigerants, components, and equipment for heating, air-conditioning, refrigerating, and related fields. This is partly due to the fact that they work for various types of companies, but also due to the fact that many people handle some selling along with other functions. For example, manufacturers, distributors, and wholesalers must all have salespeople on the road and in the offices, frequently referred to as "inside-outside salespeople."

Their prospective customers they call on (phone and/or personal visits) include all local supply stores, national distributors, regional distributors, supply wholesalers, dealer-contractors, independent service contractors, and general consumers.

SUPPLY-STORE COUNTERPEOPLE
AND OUTSIDE SALESPEOPLE

A large number of salespeople are employed as counterpeople and outside salespeople by local and regional independent supply stores. In addition, there are people employed by dealer-contractors, the plant engineers who must stock a quantity of spare parts and supplies, the independent service contractors who must carry a stock of parts and supplies on service trucks, and the supply stores that stock refrigeration and air-conditioning supplies and equipment. Each of these must have employees who devote time and effort to estimating future needs and to ordering, receiving, and storing these items. This activity is so important that careless or inefficient handling of it can result in financial problems for a business if the required parts are not there when needed.

The largest single employer of clerks, counterpeople, and outside salespeople are the owners of the approximately fifteen hundred independent refrigeration and air-conditioning supply stores that are located in every state and in several foreign countries. This does not include the vast distribution network of major manufacturers in the industry with their distributors and warehouses.

Sales jobs in the local supply stores fall into two basic categories. The first group includes clerks and counterpeople who serve the dealer-contractors, mechanics, technicians, and operation engineers who buy supplies and parts in these stores. Many of these salespeople previously worked as service and installation mechanics or technicians, so they are acquainted with most of the customers they serve as well as their needs.

The qualifications for these jobs include complete knowledge of refrigeration and air-conditioning technology, parts, supplies, catalogs, and ordering and billing procedures, plus a proven ability to deal with people.

The second group of local supply-house employees includes all outside salespeople. They cover the local territory on a regular schedule and solicit business from builders, dealer-contractors, service contractors, plant engineers, industrial purchasing agents, and other prospects.

Refrigeration and air-conditioning supply houses are important to every person who works in any branch of refrigeration and air-conditioning sales, installation, operation, and service; and everyone working in this line will become familiar with them.

The best source of information on local job opportunities in this field would be the owners, managers, and personnel of these stores. Some of these trade associations provide rather extensive training programs.

SALES JOBS

All technicians and mechanics employed by sales, installation, and service contractors are expected to be indirect salespeople for the sale of parts, supplies, fixtures, equipment, and contract service. Many employers are willing to offer some form of incentive pay for such efforts.

This approach to sales for replacement equipment and fixtures is generally successful because the technician or mechanic is in direct communication with the customer and knows when each customer should be in the market for new fixtures or may need to expand existing equipment. Incentive pay for this purpose stimulates the field servicepeople to take a greater interest in sales. Most do not give incentive pay for sales that result from a major breakdown or normal overhaul of equipment, unless the technician or mechanic was initially involved in bringing the customer to the shop.

CHAPTER 13

SHEET METAL DUCT SYSTEMS FOR HEATING AND AIR-CONDITIONING CONSTRUCTION

The term "sheet metal worker" generally implies one of two basic categories:

1. building trades sheet metal work, and
2. precision sheet metal work (close tolerance).

The first classification is naturally a part of the HVACR industry. The building trades sheet metal workers construct and install various types of ducts, which are connected to form systems through which air is passed. The various systems of ducts are referred to as heating, ventilating, and air-conditioning systems; they are needed in homes, stores, apartments, offices, schools, hotels, and in ships, airplanes, and trains. Some sheet metal workers also make and install gutters and downspouts. The following systems require sheet metal workers:

1. *Ventilating* provides circulation of air, such as in cold air registers and exhaust fans in bathrooms and kitchens. Similar systems are needed in stores, apartment and office buildings, and factories.
2. *Heating* provides warm air that is moved from the heating unit to the individual rooms or areas being heated.

3. *Air-conditioning* provides cold air that is moved from the cooling unit to the individual rooms or areas being cooled. As in some heating systems, ductwork is used.

The general classifications of the building trades sheet metal workers are as follows:

1. A shopperson makes the ductwork, which is connected to make up the entire system.
2. A fieldperson, sometimes referred to as an "outside person," installs the ductwork that has been built in the shop.
3. A general sheet metal worker is sometimes referred to as an all-around sheet metal worker and performs the functions of both the shopperson and the fieldperson. The sheet metal worker builds the ductwork, transports it to the building or job site, and then installs it.

TRAINING AVAILABLE

Some sheet metal workers follow a union apprenticeship program for their training. The apprentice program consists of being accepted as a "trainee" by an employer and attending classes one day a week (working four days a week) or one or two evenings a week (working five days a week). In some areas they are paid during their class hours the same as work hours; but in other areas, they are not paid. This is generally a four-year program, after which time the person becomes a "mechanic" and full union member. In some nonunion jobs, the area contractors are organized and have similar training programs. In other areas there is no formal training available, so the person learns purely on the job and through individual reading and studying. Even in areas organized with the union, there are also "open" (nonunion) shops.

Many local high schools, junior colleges, and adult evening programs offer courses in sheet metal ductwork layout and fabrication at the beginning, intermediate, and advanced levels.

WORK INVOLVED

To give an example of the wide range of work these people might do, the following is a list of the trade jurisdiction of the Sheet Metal Workers' International Association (union):

1. This Association has established and claims full jurisdiction over the manufacture, fabrication, assembling, handling, erection, hanging, application, adjusting, alteration, repairing, dismantling, reconditioning, testing and maintenance of all sheet metal work. All working drawings or sketches (including those taken from original architectural and engineering drawings and sketches) used in fabrication and erection; said jurisdiction to include all flat, formed in brake or press, corrugated or ribbed sheets and all rolled, drawn, pressed, extruded, stamped or spun tubing, shapes and forms of plain or protected steel, iron, tin, copper, brass, bronze, aluminum, zinc, lead, German silver, monel metal, stainless and chrome steel and any and all other alloy metals, ferrous and non-ferrous, together with all ers, straps, plates, tees, angles, channels, furrings, supports, anchors, rods, chains, clips, frames, ornaments, trimmings, grilles, registers, castings, hardware and equipment, mechanical or otherwise, regardless of gauge, weight or material when necessary or specified for use in direct connection with or incidental to the manufacture, fabrication, assembling, handling, erection, hanging, application, adjusting, alteration, repairing, dismantling, reconditioning, testing and maintenance of all sheet metal work; said jurisdiction also to include the fastening of any and all mate-

rials and equipment specified in this jurisdictional claim, whether same be applied to wood, steel, stone, brick, concrete or other types of structure, base or materials, with full jurisdiction over the making of all connections, attachments, seams and joints, whether nailed, screwed, bolted, riveted, cemented, poured, wiped, soldered, brazed, welded or otherwise fastened and attached, and all drilling and tapping in connection with or incidental thereto.

2. Any and all types of sheet metal foundation forms, wall forms, column forms, casings, mouldings, plain or corrugated domes, slab forms, flat, ribbed or corrugated sheet forms used in connection with concrete or cement construction, including sheet metal inserts to provide specified openings, also permanent column guards.

3. Any and all types of sheets, flat, formed in brake, corrugated or otherwise formed or reinforced, and all rolled, drawn, pressed, extruded, stamped or spun sheets, shapes and forms of plain or protected metal specified for use in connection with or incidental to roofing, decking, flooring, siding, water proofing, weather proofing, fire proofing, for base and support of other materials, or for ornamental or other purposes.

4. Any and all types of formed, rolled, drawn, stamped or pressed sheet metal shingles, sheet metal tile, sheet metal brick, sheet metal stone and sheet metal lumber, when specified for use as roofing, siding, water proofing, weather proofing, fire proofing or for ornamental or any other purpose.

5. Any and all sheet metal work specified for use in connection with or incidental to steeples, domes, minarets, lookouts, dormers, louvres, ridges, copings, roofing, decking, hips, valleys, gutters, outlets, roof flanges, flashings, gravel stops, leader heads, down spouts, mansards, balustrades, skylights, cornice moulding, columns, capitals, panels, pilasters, mullions, spandrils and any and all other shapes, forms and design of sheet metal work specified for use for water proofing, weather proofing, fire proofing, ornamental, decorative or display purposes, or as trim on exterior of buildings.

6. Any and all types of sheet metal buildings including hangars, garages, service stations, commercial or storage buildings, of permanent or portable design, whether manufactured, fabricated, or erected to meet specific requirements or whether constructed of standard patented units of flat, formed in brake, corrugated, rolled, drawn, or stamped sheets, shapes and forms of plain, protected or ornamental design.

7. Any and all types of sheet metal marquees, vestibule and storm door enclosures, window frames, mouldings, cornices, pilasters, mullions, panels, sills, heads, awning covers, corner posts, stops, light troughs reflectors and deflectors, bulletin boards and any and all types of sheet metal signs specified for use in connection with or incidental to display windows, building fronts, store fronts, and theatre fronts, for fire proofing, weather proofing, water proofing, ornamental or display advertising purposes.

8. Any and all types of sheet metal bill boards, bulletin boards, and sheet metal signs specified for use on the exterior and in the interior of buildings for advertising and display purposes, and any and all types of sheet metal signs and bulletin boards specified for use in connection with or incidental to the equipment and operation of theatres, hotels, hospitals, apartments, factories and other types of buildings of interior or exterior design.

9. Any and all sheet metal work used in connection with or incidental to the equipment and operation of grain elevators, mills, factories, warehouses, manufacturing plants and commercial buildings, including elevator legs and enclosures, chutes, hoppers, carriers, spirals, automatic and other conveyors, package chutes, fire apparatus and enclosures for same, pipes and fittings, dampers, machine guards, cyclones, fans, blowers, dust collecting systems, ovens and driers, heating, ventilation and air conditioning, and all other types of sheet metal work and equipment, mechanical or otherwise, in connection with or incidental to the operation thereof.

10. Any and all types of sheet metal window frames, sash, bucks, doors, frames, trim, picture moulding, frieze moulding, wire moulding, chair rail and base panels, wainscoting, mullions, pilasters, sills, permanent vestibule partitions, smoke and fire screens, portable and permanent screens and partitions for hospitals, office, commercial and factory use, toilet, shower and dressing room partitions, elevator and other types of enclosures specified for use as equipment and interior trim.

11. Any and all types of sheet metal ceilings with cornices and mouldings of plain, ornamental, enameled, glazed, or acoustic type, and any and all types of side walls, wainscoting of plain, ornamental, enameled, or glazed types, including sheet metal tile, and the application of all necessary wood or metal furring, plastic or other materials, to which they are directly applied.

12. Any and all moving picture booths and any and all sheet metal work in connection with indirect lighting systems, including side lights and foot lights in theatres, auditoriums, schools, etc.

13. Any and all types of sheet metal work specified for use in connection with or incidental to direct, indirect or other types of heating, ventilating, air conditioning and cooling systems, including risers, stacks, ducts, fittings, retrofittings, dampers, casings, recess boxes, outlets, radiator enclosures, exhausts, ventilators, frames, grilles, registers, cabinets, fans and motors, air washers, filters, air brushes, housings, air conditioning chambers, all setting and hanging of air conditioning units, unit heaters, or air-veyor systems and air handling and air treating systems, all testing and balancing systems, including air, hydronic, electrical and sound regardless of material used including all equipment and/or reinforcements in connection therewith including all smog control, air pollution and recovery systems and component parts thereof, including setting of same by any method; any and all work in connection with and/or incidental to the manufacture, fabrication, handling, erection, installation,

maintenance and repair of solar energy systems, including but not limited to residential, commercial, institutional and industrial installation; all lagging, over insulation and all duct lining; testing and balancing of all air, hydronic, electrical, and sound equipment and duct work, and any and all other sheet metal work and equipment, mechanical or otherwise, in connection with or incidental to the proper installation and operation of said systems, and all duct connections to and from same.

14. Any and all types of sheet metal work in connection with or incidental to residential work, including metal roofing and siding, gutters, downspouts, kitchen vents, bathroom vents, prefabricated fireplaces, shower enclosures, heating and air conditioning equipment and service incidental to the proper installation and operation of same.

    Any and all types of warm air furnaces, including assembling and setting-up of all cast iron parts, all stoker, gas and oil burner equipment used in connection with warm air heating, all sheet metal hoods, casings, wall stacks, smoke pipes, trunk lines, cold air intake, air chambers, vent pipes, frames, registers, dampers and regulating devices, and all other sheet metal work and equipment, mechanical or otherwise, in connection with or incidental to the proper installation and operation of same.

15. Any and all types of sheet metal work in connection with industrial work including industrial, generating, steel and aluminum, oil refining, chemical and similar type plants and all other

work in connection therewith including exhaust, smog control, air pollution and recovery systems, air-veyor systems and component parts thereof including setting of same by any method.

Any and all types of sheet metal smoke pipe, elbows, fittings and breeching for boilers, heaters and furnaces. All sheet metal lagging and jackets on engines. Any and all sheet metal drip pans, exhaust pipes, heads, safety flues, and other appliances in connection with or incidental to boilers, heaters, furnaces, engines, machinery, etc.

16. Any and all types of sheet metal furniture and equipment, lockers, shelving, library stacks, warehouse, factory and storage stacks, bins, sinks, drainboards, laboratory equipment, etc., specified for use as equipment or incidental to the operation of offices, factories, libraries, hotels, hospitals, apartments, schools, banks, public and semi-public buildings, and for general commercial use.

17. Any and all sheet metal work in connection with or incidental to the equipment and operation of kitchens in hotels, restaurants, hospitals, lunch rooms, drug stores, banks, dining cars, public and semi-public buildings, including ranges, canopies, steam tables, warming closets, sinks, drainboards, garbage chutes and incinerators, refrigerators and other sheet metal work in connection with kitchen equipment or refrigerating plants.

18. Any and all types of sheet metal work in connection with or incidental to laundry equipment and machinery, washers, clothes dryers and laundry chutes.

19. Any and all types of sheet metal work, copper-smith work and mechanical work in connection with or incidental to the manufacture, fabrication, assembling, maintenance and repair of automobiles, buses, trucks, airplanes, pontoons, dirigibles, blimps, and other type of aircraft and equipment, and any and all types of aircraft hangars.

20. Any and all types of sheet metal chandeliers, lamps and lighting fixtures, ornaments, decorations, household ware, and miscellaneous articles for use in factories and mills; any and all types of sheet metal switch boxes, cut-out boxes, panel boards, cabinets and speaking tubes.

21. Any and all types of sheet metal badges, buttons and novelties with all hard or soft soldering in connection with same by flame or other method.

22. Any and all types of sheets, tubing, pipes and fittings, used in connection with or incidental to coppersmithing work, regardless of gauge or material. The manufacture, fabrication, assembling, erection, maintenance, repair and dismantling of all said coppersmithing work, including the bending of tubes, pipes and coils and all pipe fitting in connection with or incidental thereto, and the testing of equipment when installed to insure proper operation.

23. Boats and ships, definition and duties. Manufacture, fabrication, assembling, erection, hanging, application, adjusting, alteration, repairing, dismantling, re-conditioning, testing and maintenance of all sheet metal work and coppersmithing work in connection with or incidental to building,

maintenance and repair of ships and boats, including smoke stacks, life rafts, life buoys, crow's nests, bulkheads, telegraph and speaking tubes, switch and cut-out boxes, lagging on boilers and engines, lining of all partitions, paint and lamp lockers, refrigerating compartments, battery compartments, galleys and shower baths, ventilation and kitchen equipment, ventilation piping and fittings, sheet metal lockers, sheet metal doors, sheet metal windows, steel and non-ferrous metal sheathing, sheet metal casings for housing cable, gong pull and mechanical telegraph leads, and metal lagging for machinery, boilers, pipelines, etc., sheet metal structural partitions and enclosures including pilasters, wire mesh and incidental fittings, launch and boat canopies, galley ranges, and their smoke pipes, sheet metal dresser tops, sheet metal ventilator cowls, air tanks, fuel oil tanks, battery lockers, metal furniture, sheet metal containers for handling and storing foods, paints, water, and other materials, cooking utensils, funnels, measures and similar miscellaneous articles made of sheet metal; covers with sheet lead, such articles as battery boxes, battery shelves, ice boxes and other wooden and steel parts, and items subject to corrosion; measures, marks and cuts sheet lead to size; fits and forms it about surface to be covered by heating and hammering about the edges and into corners until snug fit is obtained; making templates, forms, developing, laying out and cutting patterns, shearing, flanging, forming, bumping, rolling, spinning, punching, stamping, riveting, soldering, and all resistance welding (including, but not limited to spot and seam welding) performed on machines designed for that purpose in connection with fabrication, assembly and repair of all sheet metal and all reinforcements in connection with the above specified work.

24. The right to apply and install any and all types of slate, tile, asbestos shingle, and asphalt shingle roofing; any and all types of prepared paper and felt roofing; any and all types of sheet roll, plastic, asphalt, tar, slag, gravel or other composition roofing, specified as insulation or waterproofing in localities where there is no established local union of the United Slate, Tile and Composition Roofers Damp & Waterproof Workers Association.

25. Any and all welding in connection with the work specified in this article.

26. Railroad shopmen shall include sheet metal workers (tinners), coppersmiths and pipefitters employed in shops, yards, buildings, on passenger coaches, work equipment, etc., and on engines of all kinds, skilled in the building, erecting, assembling, installing, dismantling and maintaining parts made of sheet copper, brass, tin, zinc, white metal and lead, black planished, galvanized and pickled iron, aluminum, stainless and chrome steel, monel metal, German silver, and any other base or alloyed sheet metal. This shall include all flat, formed in brake or press, corrugated or ribbed sheets on rolled, drawn, pressed, extruded, stamped or spun shapes, tubing or forms of any sheet metal together with all

necessary or specified reinforcements, hangers, brackets, hardware and fittings, mechanical or otherwise, regardless of gauge or weight of metal when part of the operation or fabrication of parts; brazing, soft or hard solder, torch spray or hand-soldering, tinning, leading, babbitting, bending, fitting, cutting, threading, brazing, clamping, testing, connecting and disconnecting of air, water, sand, gas, oil and steam pipes and the operating of babbitt fires and pipe threading machines, oxyacetylene, thermit electric welding on work generally recognized as sheet metal workers' work. This jurisdiction includes the work performed in the Maintenance of Equipment, Maintenance of Way and all other departments of the railroad.

These jobs are available in all parts of the country, due to commercial and industrial buildings and homes needing either heat or air-conditioning, or both, plus ventilation and air filtration.

TECHNICIANS DOING SHEET METAL WORK

Technicians may specialize in installation or in maintenance and repair, or when working for a small company they may do both installation and service. They may further specialize in one type of equipment—oil burners, solar panels, or commercial refrigerators. They generally work with heating and cooling equipment or with refrigeration equipment.

Today furnace installers are called heating equipment technicians. They follow blueprints or other written instructions to install gas, electric, oil, solid-fuel, and multi-fuel systems. After putting the equipment into place, they then install fuel and water supply lines, air ducts and vents, and pumps and other components. They may connect electrical wiring and controls and check the units for correct operation.

After the installation, technicians perform the routine maintenance and repairs to keep the system operating efficiently.

EMPLOYMENT OUTLOOK

About one half of HVACR technicians works for heating and cooling contractors. The other half is employed in a wide variety of industries primarily as maintenance technicians. About 10 percent are self-employed.

Job prospects are expected to remain good. Although few HVACR technicians leave this field of work for other occupations, the number who retire will be increasing and economic growth will produce job openings.

The median salary for full-time work in 1993 was $472. The middle 50 percent earned from $356 to $596. Apprentices begin at about 50 percent of the wage rate paid to experienced workers. They generally enjoy a variety of employer-sponsored benefits like health insurance, pension plans, paid vacations and work-related training.

CHAPTER 14

SOLAR HEATING CAREERS

People currently working in the air-conditioning and refrigeration industry should consider the current impact of the uses of solar energy. Therefore, young people interested in careers in the air-conditioning industry should also consider opportunities in solar energy.

WHAT IS SOLAR ENERGY?

The basic function of solar heating and domestic hot water systems is the collection and conversion of solar radiation into usable energy. This is accomplished—in general terms—in the following manner. Solar radiation is absorbed by a collector, placed in storage as required with or without the use of a transport medium, and distributed to point of use. The performance of each operation is maintained by automatic or manual controls. An auxiliary energy system is usually available for operation, both to supplement the output provided by the solar system and to provide for the total energy demand should the solar system become inoperable.

The conversion of solar radiation to thermal energy and the use of this energy to meet all or part of a dwelling's heating and domestic hot water requirements has been the primary application of solar energy in buildings.

The parts of a solar system—collector, storage, distribution, transport, controls, and auxiliary energy—may vary widely in design, operation, and performance. They may, in fact, be one and the same element (a southfacing masonry wall can be seen as a collector, although a relatively inefficient one, which stores and then radiates or "distributes" heat directly to the building interior). They may also be arranged in numerous combinations depending on function, component compatibility, climatic conditions, required performance, site characteristics, and architectural requirements.

One of the numerous concepts presently being developed for the collection of solar radiation, the relatively simple flat plate collector, has the widest application. It consists first of an absorbing plate usually made of metal painted black to increase absorption of the sun's energy. The plate is then insulated on its underside and covered with a transparent cover sheet to trap heat within the collector and reduce convective losses from the absorber. The captured heat is removed from the absorber by means of a working fluid, generally air or water. The fluid is heated as it passes through or near the absorbing plate and then transported to points of use, or to storage, depending on energy demand.

The storage of thermal energy is the second item of importance—since there will be an energy demand during the evening or on sunless days when solar collection cannot occur. Heat is stored when the energy delivered by the sun and captured by the collector exceeds the demand at the point of use. The storage element may be as simple as a masonry floor that stores and then reradiates captured heat, or as relatively complex as a latent heat storage. In some cases, heat from the collector is transferred to storage by means of a heat exchanger (primarily in systems with a liquid working fluid). In other cases, transfer is made by direct contact of the working fluid with the storage medium (i.e., heated air passing through a rock pile).

The distribution component receives energy from the collector or storage and dispenses it at points of use. Within a building, heat is usually distributed in the form of warm air or warm water.

The controls of a solar system perform the sensing, evaluation, and response functions required to operate the system in the desired mode. For example, when heat is needed for hot water or space heat, the controls cause the hot working fluid to be delivered to whatever system is being used, exchange heat with that system's medium, and distribute the energy to the point of use.

An auxiliary energy system provides the supply of energy when stored energy is depleted due to severe weather or clouds. The auxiliary system, using conventional fuels such as oil, gas, electricity, or wood, provides the required heat until solar energy is available again.

The organization of components into solar heating and domestic hot water systems has led to two general classifications of solar systems: active and passive. The terms active and passive solar systems have not yet developed universally accepted meanings. However, each classification possesses characteristics that are distinctively different from each other. These differences significantly influence solar dwelling and system design.

An active solar system can be characterized as one in which energy resource—in addition to solar—is used for the transfer of thermal energy. This additional energy, generated on or off the site, is required for pumps, blowers, or other heat transfer medium moving devices for system operation. Generally, the collection, storage, and distribution of thermal energy is achieved by moving a transfer medium throughout the system with the assistance of pumping power.

A passive solar system, on the other hand, can be characterized as one where solar energy alone is used to transfer thermal energy. Energy other than solar is not required for pumps, blowers, or other heat transfer medium moving devices for system operation.

The major component in a passive solar system generally utilizes some form of thermal capacitance, where heat is collected, stored, and distributed to the building without additional pumping power. Collection, storage, and distribution is achieved by natural heat transfer phenomenon employing convection, radiation, and conduction, in conjunction with the use of thermal capacitance as a heat flow control mechanism.

Applications in Different Parts of the United States

Sun, wind, temperature, humidity, and many other factors shape the climate of the United States. Basic to using solar energy for heating and domestic hot water is understanding the relationship of sun, climate, and dwelling design.

The amount and type of solar radiation varies between and within climatic regions: from hot-dry climates where clear skies enable a large percentage of direct radiation to reach the ground (southwestern United States); to temperate and humid climates where up to 40 percent of the total radiation is reflected from clouds and atmospheric dust (southeastern United States); to cool climates where snow reflection from the low winter sun may result in a greater amount of incident radiation than that which occurs in warmer but cloudier climates.

As a result of these differences in the amount and type of radiation reaching a building site, as well as in climate, season, and application, the need for and the design of solar system components will vary in each locale.

THE FUTURE

The technology of solar energy is a rather new field dominated by a wide variety of acceptable performance levels subject to rapid change and new development. As solar energy grows into a major industry, there are bound to be many changes. But it appears that investing in solar energy equipment today is a wise choice. Energy costs are taking up more of the family budget each year and a solar system installed this year will pay for itself again and again in years to come. For example, a typical hot water preheating system will pay for itself in about eight years. In the next twelve years it will pay for itself another two times. So, the financial benefit is considerable, but that says nothing about the energy security a family will have while those more dependent on the utility companies will be more vulnerable to energy shortages. Solar energy indeed seems to be a smart investment.

CHAPTER 15

ENGINEERS, TECHNICIANS, AND RELATED JOBS

Engineering, just like medicine or law, is a profession. Specifically, engineers use a knowledge of mathematics and natural science, gained by education and experience, and apply it with judgment in developing new ways to efficiently and economically use the materials and forces of nature. In other words, the engineer is a builder, a problem solver, and an adventurer who believes it can be done even when told it can't.

Although many times the roles of scientist and engineer overlap, the scientist investigates the physical world and defines its fundamental laws while the engineer uses scientific research to create new products and systems of accomplishing work (and recreation). Both are team members in a technical society, working with mathematicians, architects, and other types of engineers to ensure a brighter future. In short, the scientist tells us *what is* and the engineer tells *what can be* and *how it can be done.*

ENGINEERING JOBS

The Air-Conditioning and Refrigeration Engineer

The type of engineering we are interested in is concerned with the indoor environment: heating, ventilation, air-conditioning, and refrigeration. These engineers use machines and the laws of nature to provide us with a comfortable and healthy environment not only in movie theaters, but in hospitals, sports arenas, space capsules, and offices. Their contributions have allowed cities to grow in areas always thought to be too hot and too humid and have brought us opportunities to enjoy seasonal foods all year round, resulting in better nutrition for all.

But just as new advances ensure progress, they often present us with new challenges. Not long ago, sources of energy for today's necessities seemed unlimited. Today, however, we are addressing ourselves to an energy crisis that will continue for a yet unknown period of time. The improvement and maintenance of a clean, healthy environment while still practicing energy conservation is a responsibility of engineers, particularly air-conditioning and refrigerating engineers, whose role covers about one-third of all energy used in the United States.

The explanation of how the air-conditioning system in a modern skyscraper or the refrigeration in your own home works begins with the *research engineer* and the *engineering educator.* Research engineers develop new kinds of equipment and new means of temperature control with many hours of laboratory work and testing involved. Like scientists, they publish their findings, which enables other engineers to use accurate and reliable information in actual design work. The engineering educator also uses this information in the classroom so that tomorrow's engineers have the benefit of the most up-to-date findings and the experience of all the engineers before them.

Research engineers are usually employed by manufacturing companies, government agencies, universities, and private research institutes. And even though most engineering educators have teaching responsibilities, many are consulted as experts for the solution of highly technical problems in industry and in government.

The next step in the engineering process is the work of *design engineers.* Design engineers use the work of researchers to calculate what size equipment should be used, what materials will be needed, where equipment will be located, and where air ducts, piping, and other systems will be constructed. They create the mechanical system. Their drawings or plans are eventually reproduced as blueprints and show what the mechanical system will look like after it is constructed. They also describe, in documents known as specifications, what these materials and equipment will be. Because they must carefully consider the overall building plan, design engineers work closely with architects and are usually self-employed professionals or part of a consulting firm.

Systems designed by the design engineer are made up of many components, each component being a minisystem in itself. And just as the companies that make these components employ engineers and engineering technicians in research, equipment design, and equipment testing, these companies need engineers in sales to act as consultants to the people using their equipment. *Sales engineers,* therefore, must be able to work with many different people of different backgrounds, often working with customers helping them to select the best equipment for their new needs. In order to perform these tasks well, sales engineers must be knowledgeable in business matters and must have an understanding of the people they deal with as well as the technical nature of their products.

As members of a team, all of these engineers work together to produce a system design, but it still must be built. That is the work of the *construction engineer.* The construction engineer may be employed not only by the builder but also by the design firm and by the people who will own and operate the building. These engineers ensure that the work is done according to the plans and specifications prepared by the design engineer, and they are skilled in design, operating procedures, methods of construction, and contractual law.

The highly sophisticated air-conditioning, heating, ventilating, and refrigeration systems found in today's large buildings and complexes require management by yet another engineer, the *buildings system engineer.* This engineer sees to it that systems are operating as they should and supervises regular maintenance and repair work. A full knowledge of the design and operation of every piece of equipment in a mechanical system, sometimes using the aid of computers, is required.

Education Required for These Jobs

Most air-conditioning and refrigerating engineers receive their first college degree in mechanical engineering. However, many electrical engineers enter the field by selecting to specialize in system control. Likewise, many chemical engineers become involved in refrigerating engineering when employed by certain large industries that require refrigerated processes in manufacturing.

Engineering technicians usually are graduates of two-year associate degree programs or four-year programs such as those offered by many state and community colleges. In general, engineering technicians aid the engineer in every step after the research process.

To enter or advance in certain fields or positions, advanced degrees in engineering or management are often necessary. Research engineering or a career in engineering education, for example, usually requires a Ph.D. (doctor of philosophy) degree in engineering. Engineers who decide to go into management usually find that a master's degree in business is very helpful.

For all engineers, continuing education becomes a part of life. It's essential to continually read technical journals and to take formalized courses to keep up on new developments in the chosen field of specialization. Very often, employers will pay all or part of the tuition necessary for coursework. Even while undergraduates, many engineering students find part-time work in engineering offices where they can learn and earn as engineering aides. The school of your choice may even have a "co-op" program available where you can work and be paid one semester and attend classes full time the following semester until graduation. The practical experience can be a great help to a student in making early choices about career goals.

PLAN AHEAD

You may already be preparing for an engineering career. If you are a student in junior or senior high school and find mathematics and science your favorite subjects, you should give serious consideration to an engineering career.

If you enjoy building models, working on team projects, solving puzzles, or asking why things work in a particular way, you may never be satisfied with any other career.

Because engineering is a profession, a college degree is essential. Prepare yourself now by taking college preparatory courses that include all the mathematics and sciences available. But don't neglect English or humanities studies at any time in your education because an engineer is a professional who must be able to work with and understand the problems of others and communicate well in speaking and in writing. Most importantly, keep your grades high and develop good study habits to ensure entrance into the engineering school of your choice.

Learn all you can about engineering as a career. See your guidance counselor for information on engineering and engineering schools; seek out professional engineers among friends, relatives, and neighbors; and discuss with them the kind of work done by engineers.

YOUR FUTURE

The future of engineers in the field of air-conditioning and refrigeration is indeed bright. While the profession of engineering dates back to early Egypt, this particular field of specialization didn't appear until the twentieth century. It's a rapidly advancing field that is already a basic part of life in the United States and other parts of the world. While it is a specialized field within engineering, the opportunities are broad enough within the field to allow engineers to switch their career goals several times while retaining and making use of their experience. The future and career opportunities are virtually unlimited and the chance to better human life makes it extremely fulfilling. As for personal considerations, the financial rewards of engineering are consistently better than many other four-year degree careers you could choose.

ADDITIONAL CAREER INFORMATION

The American Society of Heating, Refrigerating and Air-Conditioning Engineers (ASHRAE) is a nonprofit membership organization operated for the exclusive purpose of advancing the arts and sciences of heating, refrigeration, air-conditioning, and ventilation.

Founded in 1894, ASHRAE now has nearly forty thousand members, people who share a common interest in the many aspects of human comfort and environmental control. Approximately 10 percent of the society's membership reside in countries other than the United States and Canada, and ASHRAE is affiliated with twenty-six other organizations around the world, which are devoted to the same or to similar fields of interest.

ASHRAE also offers a student membership, which is open to anyone pursuing a course of study in a university, college, junior college, or technical institute, as well as anyone interested in the industry and sponsored by a full-grade Member or Associate Member.

You may request further career information and membership information from:

American Society of Heating, Refrigerating,
and Air-Conditioning Engineers (ASHRAE)
1791 Tullie Circle, NE
Atlanta, GA   30329–2305

CHAPTER 16

OPPORTUNITIES IN THE U.S ARMED FORCES

In response to an increasing demand for technical training, educational orientations have shifted toward career education and, concurrently, there has been a burgeoning growth of vocational training centers at both secondary and postsecondary school levels.

In addition to state-subsidized vocational training centers, many industries are sponsoring technical training for persons entering specific jobs.

Another source of training and a practical career alternative is employment in the military services. Military technical training and work experience appear to have widespread application in civilian jobs.

This emphasis is providing a stimulus to qualified young men and women to investigate the military services as a source of immediate employment and basic and advanced technical training that they can utilize in their military or civilian career planning.

The U.S. Military Enlistment Processing Command (MEP-COM) is a joint service agency that mentally and physically tests all applicants for enlistment in the armed forces, manages the Department of Defense High School Testing Program, and directs administration of the Armed Services Vocational Aptitude Battery (ASVAB).

The ASVAB measures a student's vocational aptitude and identifies those areas in which he or she is most likely to succeed in military training. The identification of the student's aptitude, coupled with the identification of specific jobs in that aptitude area, or cluster, is of considerable benefit to the student and the counselor in formulating the student's post-high-school plans.

The ASVAB is used by the armed services for recruiting purposes and also by school counselors for vocational guidance counseling.

Test scores are provided to the school counselor who in turn furnishes them to the student, the recruiting services of the armed services, and the U.S. Coast Guard.

Information about individuals who have taken the ASVAB is maintained on a computer tape for recruiting purposes for not more than two years. Scores are kept for a longer period for research purposes to assist in evaluating and updating test materials. However, personal identifying information (name, social security number, street address, telephone number) is removed from existing records after two years.

GENERAL ENLISTMENT QUALIFICATIONS

The general qualifications necessary for enlistment in the military services are listed below. Specific requirements for each service or enlistment option within a particular service vary.

Age: Between seventeen and thirty-five years. Consent of parent or legal guardian required if under eighteen. Individual services have specific requirements. Seventeen is the lower limit, thirty-five is the upper.

Citizenship: Must be a U.S. citizen or an immigrant alien lawfully admitted to the United States for permanent residence who has an immigration and naturalization form.

Physical: Must pass a physical examination. Any man or woman in normal health should have no trouble qualifying.

Marriage: Can be either single or married.

Education: High school graduation or equivalent is desired by all services and is a requirement for some enlistment options. All applicants will be tested with the Armed Services Vocational Aptitude Battery (ASVAB) to identify aptitude areas at the time of application unless valid high school ASVAB scores are available.

Specifics for each branch vary somewhat and are enumerated in the *Occupational Source Book* available on written request from:

U.S. Military Enlistment Processing Command
Building 83
Fort Sheridan, Illinois 60037

It includes information on all branches of the U.S. military, including:

Army	Air Force
Navy	Coast Guard
Marine Corps	Military Reserve Components

In addition to basic entry qualifications, this information book provides details regarding basic pay, all benefits, training programs, promotional opportunities, and the many job classifications. Further information on each specific branch of the U.S. military is available from:

Army:
Chief, Individual Training Division
Headquarters Department of the Army (DAMO-TRI)
Room 2A-712
Pentagon
Washington, DC 20301

Navy:
Head, Training Development Program Section
Office of the Chief of Naval Operations
OP-210E
Room 2815—Navy Annex
Washington, DC 20370

Marine Corps:
Deputy Director, Training Division
Headquarters U.S. Marine Corps (Code OTT)
Room 2321—Navy Annex
Washington, DC 20380

Air Force:
Chief, Training Programs Division
Headquarters U.S. Air Force (AF/MPPT)
Room ID—238
Pentagon
Washington, DC 20330

AIR-CONDITIONING AND REFRIGERATION JOBS IN THE MILITARY

These jobs are in the U. S. O. E. classification of "manufacturing." The comparable civilian job title is refrigeration mechanic. The various military classifications are:

Army:
Refrigeration mechanics

Navy:
Refrigeration and air-conditioning mechanics (MM-4294)
Shore base refrigeration and air-conditioning technician
    (UT6104)
Aviation structural mechanic, safety equipment (AME)

Air Force:
Refrigeration and air-conditioning specialist (54530)

Marine Corps:

Refrigeration mechanics (1161)

Aircraft maintenance ground support equipment refrigeration mechanics (6078)

Coast Guard:

Machinery technician (MK)

In addition, related service jobs include:

Navy:

Machinist's mate (MM)

Utilitiesman (UT)

Engineman (EN)

Air Force:

Heating systems specialist (54750)

Missile facilities technician (54150)

A basic description of the refrigeration mechanic job classification is:

1. Installs, modifies, and repairs refrigeration, air-conditioning, and ventilation equipment and systems.
2. Installs mechanical, pneumatic, electronic, and sensing/switching devices designed to control flow and temperature of air, refrigerants, or working fluids.
3. Connects wiring harnesses to electrical equipment.
4. Shapes, sizes, and connects tubing to components such as meters, valves, gauges, traps, and filtering assemblies using special bending, flaring, and coupling tools and oxyacetylene torches for soldering and brazing.
5. Conducts tests of installed equipment.

The specific desirable qualifications include:

- Courses in mechanics, machine shop, electricity, and practical mathematics.
- Mechanical aptitude. Experience in garage, power plant, or machine shop.

Annual Earnings

In 1992 the average compensation of all military personnel was $27,970, which included basic pay plus housing and subsistence allowances. Enlisted personnel averaged $24,280, warrant officers $40,500, and commissioned officers $50,400. There are also advantageous benefits and educational opportunities in the military.

CHAPTER 17

YOUR OWN BUSINESS

Many skilled tradespeople have considered starting their own businesses, especially if they have worked for someone who started a business recently and did well. But remember that the best tradesperson is not necessarily the best businessperson. To successfully operate a business, you must be able to manage, make good estimates, sell, and coordinate work and employees. You must also have adequate working capital and loan sources. As soon as you start a business, you have many more responsibilities than the tradesperson—responsibilities to your new business, to your employee or employees, to your customers, and to your family. This chapter is basically aimed toward supplying the prospective owner with some basic information to help him or her avoid the pitfalls and to forge ahead to success.

WORDS OF CAUTION

Many people with very small contracting businesses hire one or two employees in addition to working themselves, and they merely make a living equal to their wages if employed by someone else. But if business is good, they can work additional hours each week and do better financially. If they are able to grow and hire three to five employees, they are able to gradually increase their profits as their volume of business increases. As they continue to grow, their profits should also continue to increase. Although the owner is actually making a higher hourly rate for each hour worked, additional skills not required as a tradesperson are learned and utilized—the owner is a manager, estimator, salesperson, coordinator, and investor. As business continues to grow and employees assume some of these responsibilities, the owner's responsibilities in management and overall coordination become even greater. If possible, the businessperson is able to gradually devote fewer hours per day or week to the details of the business, which is one of the goals or incentives for starting your own business. But remember that getting to this point requires an above average amount of determination and management ability.

Risks as well as opportunities have to be considered, though, and this chapter will show you ways you can minimize your risks and improve your probability of success. You can certainly learn and profit from others' mistakes and successes.

DECIDING TO OWN A BUSINESS

Before a person can decide whether to own a business, the advantages of owning a company should be considered, chances of success should be studied, and qualifications—including health—should be examined. If a person finds that his or her reasons for wanting to own a company are not sufficient to risk failure, the candidate can decide against business ownership. Or, if someone believes he or she is not qualified to own a company, that person can continue to work as an employee rather than an owner. But when a person has good reasons for wanting to own a company and believes that his or her qualifications promise a good chance of success, financial position can be improved.

REWARDS OF OWNING A BUSINESS

One of the primary reasons for owning a business is the opportunity for financial gains. The profits from a well-managed, prosperous business can provide a tremendous income for the owner. For all practical purposes, there is no limit to the income that might be provided by a company. This is in sharp contrast to the typical employee's salary, which has definite limits.

But not every business owner expects to get rich—and it is a mistake to assume that every entrepreneur is trying to amass a personal fortune. Some people go into business with the realization that their income will not be substantial. Consider, for example, the owner of a service station, small bakery, or small automobile repair shop who may earn only a reasonable income. The income might still be higher than the person would earn as an employee. In other words, an individual may have financial reasons for owning his or her own business even if that person believes the monetary rewards will not be great.

Financial reasons, as important as they are to businesspeople, are not the only incentives for owning a company. The desire to be independent—"to be your own boss"—is another reason that leads many people to business ownership. Some people are simply unhappy when they are subject to authority and control. Unfortunately, a person who works in a large organization sometimes has little free time and perhaps only a small degree of independence. But it is important to realize that while the owner of a small business has much greater independence than a typical employee, an owner is still not perfectly free to follow his or her own desires. A small business owner must work with and generally satisfy customers, suppliers, employees, the government, and even competitors to some extent. A business owner who neglects responsibilities to these groups is not likely to remain in business long.

The owner of a small business is also likely to be motivated by a sense of pride and satisfaction. A person who has the drive, initiative, and ambition to own a business is generally respected in the business community and can readily measure his or her accomplishments. Employees, on the other hand, usually cannot be as proud of the business, and they cannot always see the fruits of their labor.

YOUR CHANCES OF SUCCESS

One of the great dangers in owning your own company is the fact that young, small companies have a high rate of failure. The majority of business failures are firms that are young (five years or younger) and small. Over 80 percent of the failing firms have liabilities under $100,000. Obviously, a person who starts a business or buys an existing company must be willing to accept the possibility of failure. A businessperson, however, might be able to minimize the chance of failure by being familiar with the failure rates of various types of firms. It is also helpful to know that failure rates differ considerably in different parts of the country. As a general rule, failure rates are lower in the South and Midwest and higher in the western part of the United States. Specifically, the rates of failure have recently been high in California, Arizona, and Oregon. This is probably as a result of rapid population growth and extreme competition.

Failure rates are also much higher during some years than others. Of course, many companies failed during the Great Depression of the 1930s, but the rate of failures was also high in 1961 and a rash of failures developed in 1970. This means that a person who is considering the possibility of owning a business should give some thought to economic conditions.

COMMON CAUSES OF BUSINESS FAILURES

What causes business failures? In view of the high rate of failures of small businesses, this question deserves the serious attention of anyone who is contemplating business ownership. It is helpful to know that the apparent causes of failure can be traced to underlying causes. According to recent studies, the basic causes of business failures (contrary to common belief) are seldom attributable to business recessions, fraud, or disasters. Almost 90 percent of all business failures result from lack of business experience, unbalanced experience, and incompetence (or poor management and inadequate experience).

The pitfalls in managing a small business have been set forth specifically in a recent study by Dun and Bradstreet. This study states the leading pitfalls as follows:

- lack of experience
- lack of money
- wrong location
- inventory mismanagement
- too much capital going into fixed assets
- poor credit granting practices
- taking too much money out of the business
- unplanned expansion
- having the wrong attitudes

Since these pitfalls can be traced to poor management, it is obvious that management is the key for a successful small company.

IMPORTANT QUALIFICATIONS FOR BUSINESS OWNERSHIP

It should be clear now that not everyone should attempt to own a business. The owner of a small business increases the chances of success by meeting four major qualifications:

1. appropriate personal characteristics
2. proper experience and education
3. an adequate degree of management ability
4. sufficient financial capacity

Exactly what does each of these terms mean to you? The following paragraphs explain them.

Personal Characteristics

A proprietor who can get along well with customers, employees, and business associates can achieve greater success. The problem is that it is hard for us to objectively evaluate our own personality. A business owner needs a high degree of drive, ambition, and willingness to work long hours. Some people who would enjoy the freedom and independence that are associated with business ownership are just not capable of forcing themselves to work.

Education and Experience

You have probably heard stories about the "poor kid" who had to quit school at an early age and later developed a multimillion dollar business. But you should realize that it was easier in the past to become a business success without an education than it will be in the future. Education plays an important role in determining a person's occupation, and it would be a mistake to believe that a self-employed person does not need an education. Indeed, since the owner of a business usually must manage the entire business operation, there is a greater need for education than for an employee who works at a specialized job. In the past a high school education, at least, was a necessity for business ownership. Now junior college or at least some college-level courses will help put a person in a much better position due to the increased competition in business today. The technical knowledge of your line of work isn't enough.

But few people are able to handle a company, even a small company, immediately after they graduate from school. Experience is important, but experience—by itself—does not necessarily prove that you can manage a business. The kinds of experience you have had are critical. It is not enough to have extensive experience in one specialized area. For purposes of owning a business, it is more valuable to have five years of varied experiences than to have one year's experience five times (performing the same limited duties all five years). Before starting your own business, get some experience in estimating and selling, in addition to knowing all facets of your specific trade. Also try to learn a little about the office procedures necessary for a contracting business.

Management Ability

Many people know the operation of a business very well and yet they cannot manage their own company. At the same time, many of the most successful companies are managed by people who know practically nothing about the day-to-day operations of their specific trade or line of work. Sometimes the ability to manage is fairly separate and distinct from education and experience.

An effective manager must have the ability to plan, organize, direct, and control a company. Normally this requires a good general knowledge of business practices and, perhaps more importantly, an ability to work with others. Every businessperson must work with other people, and as a business grows, a manager must be more and more able to delegate responsibility to others. Many small businesses are stifled because the owner only lets the business grow to the extent that one person can handle the day-to-day work.

BASIC EQUIPMENT NEEDS FOR A CONTRACTOR-SERVICEPERSON

The contractor-serviceperson who starts his or her own shop needs certain tools and equipment, the most expensive item being the service truck. It does not need to be new or fancy, but it should be neat in appearance, reliable, and economical to operate. It should also have storage compartments that can be locked for safekeeping tools and supplies. Most beginning contractor-servicepeople have acquired most of the necessary tools and equipment while working for others. Don't invest in a lot of expensive tools that may never be used. The following list includes most of the basic needs:

- a set of mechanic's hand tools
- a reliable leak detector
- a test manifold, including gauges and hoses
- a good portable vacuum pump
- an electric drill motor, with drill bits, and hole saws
- a complete acetylene welding outfit (small tanks preferred)
- a reliable test meter for electrical circuits
- cylinders for commonly used refrigerants
- a small stock of brass, copper, and steel fittings and copper tubing
- printed forms for service calls and receipts

The contractor-serviceperson has the responsibility for ordering parts, supplies, equipment, and refrigerants; and one of the most important elements in the successful operation of this business is the relationship developed with refrigeration and air-conditioning supply houses. If these suppliers have local stores, the problem is simplified. However, every contractor-serviceperson should maintain the friendship and goodwill of supply store owners, managers, and sales personnel.

The local supply store is often a meeting place for the people of the local refrigeration industry and a clearinghouse for much technical information and industry gossip. This form of communication has social value and provides many good tips on jobs and new business.

RAISING ADEQUATE CAPITAL

As we have seen, many would-be entrepreneurs (owners of a business) are stymied by the problems of raising enough money to get started. Many people undoubtedly give up before they really know if they can raise enough money. Actually the only way a person can determine whether it is possible to raise sufficient capital is to study the need for funds and investigate the possible sources. Funds may be needed for rent, inventory, machinery, equipment, salaries, taxes, insurance premiums, and legal fees. When the total expense of starting a business is calculated, it usually is a surprisingly large amount. This is probably one of the major reasons why many people decide they should not own a business. Most people who meet the other qualifications reasonably well can probably raise enough money to get a business started. In addition to a prospective business owner's money, funds may be raised for a promising business venture from many different sources, as discussed later.

A small business usually requires four different types of funds. The first is *working capital*. Businesspeople normally use this term to mean total current assets including cash and inventory in order to operate. The second type of capital needed is *fixed assets capital,* which are the long-term, reasonably permanent assets that are used in the business. Unlike current assets, they are not normally sold or converted into cash. The most common examples of long-term fixed assets are land, building, machinery, equipment, and automobiles or trucks. Often a small business can reduce its needs for fixed asset capital by renting or leasing.

Initial expense capital is usually needed by a small business. This includes funds for legal fees, state organizational taxes, expenses of finding an appropriate location, and advertising costs. These are the start-up costs that are not included in the other needs for funds.

In most small businesses, there is a fourth need for capital—*funds for personal expenses.* Although this is technically not a part of the business capital, as a practical matter most new business owners need money for personal living expenses while the company is getting started. Many companies cannot immediately operate at a profit, even though they will be able to generate profits after the company is firmly established.

Every company is unique to some extent, and therefore the needs for capital depend on the characteristics of each company. Furthermore, the need for working capital may be seasonal. Consequently, when capital needs are projected into the future, a businessperson cannot always get a precise answer and may only arrive at a reasonable estimate of capital needs.

After arriving at an estimate of capital needs, a person must investigate the sources available for the needed capital. A new business can be financed in a number of ways, and a person who is interested in owning a business should explore all possibilities. Personal savings are the most important source of funds for a small business. It is almost impossible to start a business with no financial contribution from the owner. But if a person is willing to put up personal savings, it may be possible also to obtain funds from others. Sources of loans include:

- loans from relatives and friends
- loans from previous owners, if purchasing an established business
- commercial bank loans
- trade credit (loans made by suppliers)
- equipment supplier loans
- life insurance

SIZE OF BUSINESS

Businesses classified under the *construction* heading tend to be smaller with over half having one to three employees. So the necessity of starting small is not a difficulty in the construction industry. In fact, only 3.2 percent of the construction businesses hire fifty or more employees.

BUYING AN EXISTING BUSINESS

The advantage of buying an existing concern is that it might be less risky than starting a new company. When a new company is started, the sales forecast (no matter how carefully constructed) may not be very reliable and, consequently, the profit earned by the owner of a new company is largely unpredictable. But an existing business has an actual operating record that can be studied. As a result, sales and profits are much more predictable. When the ownership of a company changes, the newcomer may not be able to do as well as the previous owner—or he or she may be able to do better. But in any event, a person who buys an existing business should have some idea of how successful the company will be.

Another important reason why there is less risk associated with an existing business is that a going concern has already established a group of customers, and often the clientele will remain loyal to the business even after it has changed ownership. In addition, an existing company may have valuable relationships with others, such as employees, suppliers, general contractors, and union officials. Apart from the idea of risk there is another major advantage of buying an existing company rather than starting a new one. The time, effort, and cost of starting a new business may be avoided or minimized by buying a going concern. Of course, time and effort must be devoted to financial and legal problems when a company is purchased, but the problems should be fewer because an existing company in good standing normally has inventory, physical facilities, employees, and other necessities.

If you become interested in buying a business and you discover one for sale, one of the first logical questions you should ask yourself is, "Why is the present owner interested in selling the business?" This is not only a logical question, but an important one. The seller may tell you that he or she is retiring because of poor health, old age, or a desire to live elsewhere. But these may not be the real reasons for selling, or they may be only part of the true reasons. The owner may be having trouble from competition, rising costs, increasing taxes, or labor problems. Since there is always the possibility that the owner has hidden reasons for selling, a prospective buyer should always probe carefully into the business for possible problems. This may be done by talking to competitors, employees, suppliers, and others who may know something about the business.

A buyer should insist upon sales and profit figures for at least three preceding years. If the owner will not supply these figures, a buyer should be cautious. If the owner has the figures and will not make them available, the buyer should suspect a major problem. And if the owner cannot show business figures because he or she has not kept adequate records, the buyer should realize that the business firm may be less than genuine.

In analyzing the company's financial statements, a buyer should study the company's sales figures, expenses, profits, working capital, inventory turnover, and contingent liabilities. In these studies, he or she should investigate the quality of the company's assets and determine whether they are valued accurately. The same applies to business liabilities.

In most cases, it would pay the buyer to hire an independent auditing firm for a professional financial analysis of the company. If the company appears to be a wise purchase, it is always advisable to have an attorney check the company's contracts and help transfer the company to you.

STARTING A NEW BUSINESS

There are three major reasons why many people prefer to start their own company rather than to buy a going concern. First, if you start your own company you can select your own location, inventory, employees, bankers, and products. In other words, you can avoid the mistakes of previous owners. There will be more risk in a new company, but also greater opportunity if you manage the company well.

A second reason, which is similar to the first, is that it is sometimes difficult to find a suitable existing company. It is easy to find companies for sale, but there are probably more unsuccessful than successful companies for sale. As a general rule you can buy a poor company cheaply or a good company at a high price. This dilemma naturally leads many people to start their own business.

Sometimes it is absolutely necessary to start a new company if you want to go into business for yourself. This is the case when you want to sell a product or service that is currently not available in your geographic area.

Starting a new company, as we have seen, involves a considerable amount of risk. Ideally, a person should only start a new company after he or she has investigated and planned carefully. Some of the problems of starting a new business have already been discussed, such as raising capital and selecting the proper form of organization. But, in addition, a person who starts a new business should be convinced that there is a genuine need for his or her product or service and that these needs can be met effectively and economically. If the product or service is already being provided by other companies, a new company should not be started unless there is evidence that the demand for the product or service has expanded or existing companies are not being managed effectively. In other words, a company should not be started simply "to get into business." And if a new company is organized to sell a new product or service, a prospective businessperson should realize that competition may be fierce if the product or service is well accepted.

SELECTING A LOCATION

In selecting a location, you must decide on the specific town or city, the area within the town, and the specific site within the area. You have probably already selected the town or city where you currently live or desire to live. Although it is important to select a place in which you and your family will be happy, you must make sure the community needs the business you plan to open. Consider the population and the current and anticipated growth rate. Also consider the age, occupations, and income breakdown of the area. Is there a definite demand for your product or service? Check carefully the number of competitors and their sizes to determine whether your business is needed. Do some chain, department, or discount stores provide any competition for you? Does a local lumberyard also have a remodeling department or other maintenance services?

The larger the town the more important the selection of area within the town becomes. People generally patronize businesses within their immediate neighborhood, so you must be careful to select an area where the residents need and can afford your product or service and where there is not already too much competition. In some areas, one well-established business might already be adequate and you won't have a chance. This might be true if the established business has had a long and good reputation with a well-known and liked owner. Therefore it is possible that the residents wouldn't give a new person a chance. In another area, three relatively new and so far unreliable competitors might provide you a very good opportunity for success. The residents are currently dissatisfied and willing to give you a chance. So be sure to check the size and reputation of your competitors.

Depending on the type of business you are starting, the specific site or location within an area might not be very important, especially if most of your business is done with general contractors or by telephone calls from potential customers. In such a case, most important to you is a building and land adequate for your operation and anticipated growth within a reasonable length of time. Don't sign a lease and rent facilities that you will probably outgrow in six months. On the other hand, don't rent facilities so large that you will probably not need all the space until five years in the future. Naturally, the determination of the best size of building requires some "guesstimating" when you are just starting your new business. Most important, never overextend yourself or be optimistic beyond reality.

When you find a building, be sure it has adequate office, shop, storage, loading, and parking space. Check into its recent history. If it has experienced frequent change, find out the types of businesses that were there and why they left. If it has been vacant for a while, try to determine the reason. Then get the most advantageous lease you can for yourself. Don't sign for too long since you hope to outgrow the facilities. But do try to include a renewable clause stating the rent so the owner doesn't unfairly raise your rent as he or she sees you becoming successful. Try to sign the lease under your business name, not personal name, so you are not responsible if your business fails.

Although not necessary for most contracting businesses, you might want to consider a location in or near the shopping district. If so, fully utilize your window space and make sure your company name is prominently displayed. If such a location is not too expensive, the familiarity might help. Every time people walk or drive past your business, they are reminded of your company name and think of it first when they need your product or service. But don't overestimate the value of such a location, for in some types of construction or building businesses it is of no value.

MANAGING YOUR BUSINESS

Management of a new small business involves many details. There is much more to running a business successfully than seems obvious at first glance. A successful owner must be able to plan, organize, direct, and control the business. A successful owner-manager is a person who performs all of these functions effectively, but owners do not always realize their management functions and how important they are. When the owner of a small business gets too involved in day-to-day work and neglects management functions, business is bound to suffer.

Strive for a Balanced Business

You will probably get your start with just one type of business such as new work, modernization, replacements, or service (depending on your specific type of business), usually limited either to residential, commercial, or industrial work. But, as quickly as possible, try to diversify so you cover more than one type of work within your trade area. In this way you will be able to somewhat overcome the seasonal and cyclical slow times. In many lines of work, a well-developed service business provides a dependable foundation since service business is always available. A well-balanced business is a key to continued success.

Traps to Avoid

Since so many failures result from poor management, a closer look at these problems can help guide us properly. Basically the traps to avoid include:

1. *Inadequate records.* Among the other things, good records help when estimating and submitting the bid for each job. Due to the close competition in many lines of work today, an accurate estimate is essential. If you are too low, you will get the job but not be able to make a profit. If you are too high, you will not get the job.
2. *Cumulative losses.* Small losses on too many jobs add up to a serious loss.
3. *Tax requirements.* Remember these obligations both when making estimates and when budgeting your monthly payments due.
4. *Growth problems.* The biggest neglect when a business is growing is in the bookkeeping system. It must grow with you and a little money spent on new forms and even professional advice can save you a big problem later.
5. *Cost analysis.* Keep a record of the way each employee spends each hour. Don't charge shop clean-up and improvements to only one specific job because this distorts the records for future reference.

6. *New product developments.* Keep alert to new developments both in products and in work methods. Sometimes a new product or machine can save so many labor hours that its higher cost is actually more economical in the long run.

7. *Product and market diversification.* If at all possible, don't limit your work to a single product or service or to just residential, commercial, or industrial jobs.

8. *Check credit and collection policies.* Keep a system for your collection policy and for checking credit references before granting any credit.

9. *False confidence.* Don't expand until you have the sales to justify it.

10. *Legal problems.* Saving money by avoiding legal advice or assistance is really not a savings because the problem seldom goes away by itself. When it has grown into a larger problem the legal fees have also increased.

11. *Family factors.* Don't let family obligations hinder your success, especially by employing relatives who really can't handle the job you need to have done.

12. *Administrative coordination.* You must be able to coordinate sales and completion dates promised.

13. *Technical knowledge.* Don't venture into something you have no background knowledge about and cannot learn quickly enough.

14. *Internal conflict.* Problems among partners or among employees can eliminate your profits.

A key to success is to be able to look and think ahead and offset potential trouble before it becomes serious.

Importance of Self-Improvement

No matter how good we think we can manage our business, we can always improve. The person who is modest enough to admit this is actually a big step ahead toward success and greater success. A nearby community college or chamber of commerce probably offers some courses or seminars for small business owners. When you see the outline of the course content, it might seem too elementary for you; but remember that it is the group discussions of mutual problems that are the most valuable and worthwhile part of the course. Try to find a course that is specifically organized and offered for businesspeople, not for college students. If none is available, then take a course with college students and you will be surprised how much you can learn. Consider courses such as introduction to business, accounting, sales, marketing, or introduction to management or office management.

THE SMALL BUSINESS ADMINISTRATION

The Small Business Administration (SBA) is part of the U.S. government and is designed to help small businesses. It does this in several basic ways, including the distribution of printed materials, advice and counseling services, government contracts advice, and loans to small businesses. You can write for the printed materials concerning each of these services by addressing your correspondence to:

Administrator
Small Business Administration
Washington, DC 20402

Be sure to request the administration's lists of free or inexpensive publications. Or if you can find a regional SBA office listed in your telephone directory, either write or call that office for the same information.

SAMPLE COURSES OF STUDY IN HVACR

Since there is a variety of types of instruction available in air-conditioning and refrigeration, this section includes a sample curriculum in each:

- associate degree program (two-year college degree)
- private trade school
- correspondence courses

ASSOCIATE DEGREE PROGRAM

This is a two-year college program typical of those offered at many junior colleges and community colleges. The sample below is from the College of Lake County, Grayslake, IL. Following it are some course descriptions.

Refrigeration and Air-Conditioning
(Associate in Applied Science)

This program provides instruction in air-conditioning, heating, and refrigeration. Introductory courses in electricity, electric motors, and theory of refrigeration are included. Advanced work in the commercial area includes work on reach-in and walk-in units found in stores, dairies, and markets. Other areas of study include uses of air-conditioning, temperature and humidity control, air circulation, cleaning, installation, and troubleshooting of equipment.

First Semester

RAC	110	Theory of Refrigeration	5
RAC	174	Applied Electricity	4
MTH	115	Applied Mathematics II	3
ENG	121	English Composition I	3
			15

Second Semester

RAC	113	Commercial Refrigeration Systems	4
RAC	119	Electric Motors and Controls	5
		Social Science Elective.	3
RAC	112	Residential AC Systems	4
			16

Third Semester

RAC	118	Residential Heating Systems	4
RAC	114	Commercial AC Systems	4
		Technical Elective	2–4
PHY	111	Technical Physics I	4
			14–16

Fourth Semester

RAC	173	Air Movement and Ventilation	4
SPE	111	Communications II	3
RAC	117	Installation and Service Problems .	4
		Humanistic Studies Elective.	3
ECO	110	Economics for Business and Industry	3
			17

Total Hours   62–64

REFRIGERATION AND AIR-CONDITIONING

The two certificates allow students to specialize in heating and air-conditioning or refrigeration and air-conditioning. Both certificates require introductory courses in electricity, motors and controls, and theory of refrigeration system operation.

HEATING AND AIR-CONDITIONING

(Certificate)
Code 24RG

RAC	110	Theory of Refrigeration	5
RAC	174	Applied Electricity	4
RAC	118	Residential Heating Systems	4
RAC	119	Electric Motors and Controls	5
RAC	173	Air Movement and Ventilation	4
RAC	115	Installation and Service Practices for Heating and Air-Conditioning	4
		Technical Electives	8
		Total Hours	34

REFRIGERATION AND AIR-CONDITIONING

(Certificate)
Code 24RF

RAC	110	Theory of Refrigeration	5
RAC	174	Applied Electricity	4
RAC	113	Commercial Refrigeration Systems	4
RAC	119	Electric Motors and Controls	5
RAC	117	Installation and Service Problems .	4
		Technical Electives	12
		Total Hours	34

Course Descriptions

RAC 110   Theory of Refrigeration   5 hours
Lectures, demonstrations and lab assignments in the area of basic refrigeration, theory and practice. The functioning and operating characteristics of the mechanical refrigeration system including condensers, evaporators, compressors, refrigerant control devices, refrigerants, test equipment and special service procedures connected with the basic refrigeration cycle will be covered. The student will be required to provide basic hand tools that will be used in this and other refrigeration and air-conditioning courses.
Prerequisite: None

RAC 111   Domestic Refrigeration Systems   4 hours
Service needs of the domestic refrigeration industry including servicing of domestic refrigerators, freezes, ice makers, etc., covered. Various types of electric controls including thermostats, defrost controls, relays, and protective devices are studied. System malfunction diagnosis and corrective procedures are presented and practiced. The student will be required to provide their own basic tools.
Prerequisite: RAC 110, RAC 174

RAC 112   Residential Air-Conditioning Systems   4 hours
A study of the basic principles, practices, and operations of air-conditioning equipment used for residential cooling. Laboratory work includes the installation, operating, and testing and trouble-shooting of various types of air-conditioning equipment. Students will be required to provide their own basic tools.
Prerequisite: RAC 110, RAC 174

RAC 113   Commercial Refrigeration Systems   4 hours
Various types of installations are studied, along with the product to be cooled, the desired temperature to be maintained, and humidity conditions. Problems involving system balance and com-

ponent capacity and use of heat load charts are presented. Students will be required to provide their own basic tools.
Prerequisite: RAC 110, RAC 174

RAC 114   Commercial Air-Conditioning Systems   4 hours
Special attention is given to the cooling and heating requirements for various commercial structures and the selection of equipment to meet these needs. Calculations and problems coordinated with laboratory operations, heat gain, heat loss calculation, and humidification and dehumidification are included. Students will be required to provide their own basic tools.
Prerequisite: RAC 110, MTH 115

RAC 115   Installation and Service Practice for
Heating and Air-Conditioning   4 hours
Provides experiences in the installation and service of residential and commercial heating and air-conditioning equipment including selection, layout, troubleshooting, and code requirements. Students will be required to provide their own basic tools.
Prerequisite: RAC 112, RAC 118, RAC 119

RAC 117   Refrigeration Installation and
Service Problems   4 hours
Installation procedures and service techniques used in commercial refrigeration and air-conditioning, including piping techniques, codes, preventive maintenance, multiple systems, and system accessories. Students will be required to provide their own basic tools.
Prerequisite: RAC 110, RAC 113, RAC 119

RAC 118   Residential Heating Systems   4 hours
Oil burners, high pressure and vaporizing; electric heat, various types including panels, baseboards, valance, and electric furnaces; heat pumps, gas heat, installation, and servicing. Students will be required to provide their own basic tools.
Prerequisite: RAC 110, RAC 174

RAC 119   Electric Motors and Controls   5 hours
Provides a background in the theory of operations application and installation of electric control circuits and control devices used in refrigeration, heating, and air-conditioning. Covers the basic types of motors used in the industry, their operation and application. Students will be required to provide their own basic tools.
Prerequisite: RAC 110, RAC 174

RAC 171   Refrigeration and Air-Conditioning Code   3 hours
Offers students an opportunity to learn the requirements placed on contractors and installation personnel involved in the layout and installation of major refrigeration, heating, and air-conditioning equipment and will attempt to cover national, state, and local codes that govern such installations.
Prerequisite: RAC 110

RAC 172   Special Problems in Refrigeration and
*   Air-Conditioning   1–3 hours*
Individual research and projects in the area of a student's interest, involving significant effort in problem analysis, data collection, and the development of appropriate solutions. Also offered to groups if significant interest exists in specific areas such as solar energy, energy conservation, etc. Hours or credit would be arranged with instructor. Students will be required to provide their own basic tools.
Prerequisite: RAC 110, RAC 119

RAC 173   Air Movement and Ventilation   4 hours
Proper methods and techniques involved in the design, sizing, and balancing of complete ventilation systems covered. Also covers special instruments used to measure air properties and air movement. Students will be required to provide their own basic tools.
Prerequisite: RAC 110, RAC 112

RAC 174   Applied Electricity   4 hours
Basic AC and DC circuitry, laws of electricity, uses of meters, and safety procedures are included in the course. Emphasis is placed

on application of electrical wiring to heating, refrigeration, and air-conditioning. Practical techniques in wiring and parts of National Electrical Code are studied. Students will be required to provide their own basic tools.
Prerequisite: None

RAC 175   Pneumatic Control Systems   4 hours
Provides a background in the theory of operation, application, and installation of pneumatic control circuits and control devices used in heating and air-conditioning. Also covers electrical devices used in conjunction with pneumatic controls. Students will be required to provide their own basic tools.
Prerequisite: RAC 119, RAC 114

PRIVATE TRADE SCHOOLS

There are many private trade schools with full-time and part-time courses ranging from six weeks to one year, depending upon the depth of training desired. The samples here are six-week day programs or eight-week evening programs. They are available at Coyne American Institute, 1235 West Fullerton Avenue, Chicago, IL 60614.

Air-Conditioning, Refrigeration, and Heating

New concerns about energy efficiency and energy-saving operation of equipment underscore the continuing demand for skilled personnel in this growing service industry.

This training is designed for men and women with mechanical aptitude who enjoy working with their hands. Students study in six main modules: (1) Introduction to Mechanical Servicing; (2) Introduction to Electrical Servicing; (3) Introduction to Commercial Controls; (4) Commercial Controls and Applications; (5) Gas and Oil Heating; and (6) Electric Heat and Heat Pumps.

EDUCATIONAL OBJECTIVE

The day course in Air-Conditioning, Refrigeration, and Heating covers the basic and advanced theory and laboratory practice required to provide the successful graduate with the necessary skills for entry-level positions in companies that install, operate, troubleshoot, and service commercial and domestic air-conditioning, refrigeration, and heating equipment in private homes, office buildings, factories, and other facilities.

Day course classes are offered Monday through Friday, 8:00 A.M. to 1:00 P.M., 5 hours per day, 25 hours per week. The day program consists of six 6-week modules for a total of 36 weeks, 900 hours. The employment objective is Entry-Level AirConditioning, Refrigeration, and Heating Mechanic.

The evening course in Air-Conditioning, Refrigeration, and Heating covers the basic theory and laboratory practice required to provide the successful graduate with the necessary skills for entry-level positions assisting in companies that install, operate, troubleshoot, and service commercial and domestic air-conditioning, refrigeration, and heating equipment in private homes, office buildings, factories, and other facilities.

Evening course classes are offered Monday, Tuesday, and Thursday evenings, 6:00 P.M. to 10:30 P.M., 4.5 hours per day, 13.5 hours per week. The evening program consists of six 8-week modules for a total of 48 weeks, 648 hours. The employment objective is Entry-Level Basic Air-Conditioning, Refrigeration, and Heating Assistant.

COURSE OUTLINE

Introduction to Mechanical Refrigeration Systems
(6 weeks days or 8 weeks evenings)
    Nature and Effects of Refrigeration
    Mechanical Refrigeration Cycle
    Soldering and Brazing
    Flaring and Swaging

Compressors
Condensors
Metering Devices
Evaporators
Receivers
Using a Pressure-Temperature Chart
Charging Systems
Evacuating Systems
Discharging Systems
Gauges and Other Test Equipment
Common Refrigerants
Troubleshooting Mechanical Refrigeration Systems

Introduction to Electrical Servicing
(6 weeks days or 8 weeks evenings)
Basic Electricity and Motors
Electric Control Circuits
Wiring Thermostats and Pressure Controls
Introduction to Hermetic Units
Domestic Hermetic Systems
Domestic Hermetic Servicing
General Service Procedures
Troubleshooting Hermetic Systems

Introduction to Commercial Controls
(6 weeks days or 8 weeks evenings)
Domestic Refrigerators and Freezers
240V Compressor Wiring
240V Schematics
3-Phase Transformers and Supply
3-Phase Motors and Compressors
Commercial Operating and Limit Controls
Motor Starters and Contactors

Commercial Electric Defrost
Commercial Hot Gas Defrost

Commercial Controls and Applications
Water Cooled Units
Introduction to Chillers
Psychrometrics
Using a Sling Psychrometer
Introduction to Heat Loads
Basic Heat Load Calculation for Estimating Purposes
Ice Machine Theory
Ice Machine Servicing
Troubleshooting Ice Machines

Gas
Natural Gas Heating Systems
Troubleshooting Basic Heating Circuits
Spark Ignition Circuits

Heating
Air Distribution
Hot Water Heating
Steam Heating
Reading Electric Schematic Diagrams

Air-Conditioning, Electric Heat, and Heat Pumps
Basic Air-Conditioning
Central Air-Conditioning
Troubleshooting Air-Conditioning Circuits
Troubleshooting Refrigeration Systems
Electric Heating Systems
Electric Heating Controls
Heat Pump Systems
Heat Pump Controls
Rooftop Units

SCHOOLS OFFERING A COMPLETE CURRICULUM IN AIR-CONDITIONING

There are many schools in the United States offering complete courses of study in air-conditioning and/or refrigeration. A list of schools is available from:

Air Conditioning and Refrigeration Institution (ARI)
4301 North Fairfax Drive, Suite 425
Arlington, VA 22203

Its directory provides the following classifications of schools:

- Secondary: including high schools, trade schools, and vocational-technical centers
- Postsecondary: including vocational-technical, two-year colleges, and four-year colleges
- Adult education: including evening classes, continuing education programs, and life-long learning programs

Also, the Refrigeration Service Engineers Society offers ten seventy-two-hour courses at local chapters. For the nearest chapter, write to:

RSES International Headquarters
1666 Rand Road
Des Plaines, IL 60016

A complete list of titles in our extensive *Opportunities* series

OPPORTUNITIES IN

Accounting
Acting
Advertising
Aerospace
Airline
Animal & Pet Care
Architecture
Automotive Service
Banking
Beauty Culture
Biological Sciences
Biotechnology
Broadcasting
Building Construction Trades
Business Communication
Business Management
Cable Television
CAD/CAM
Carpentry
Chemistry
Child Care
Chiropractic
Civil Engineering
Cleaning Service
Commercial Art & Graphic Design
Computer Maintenance
Computer Science
Counseling & Development
Crafts
Culinary
Customer Service
Data Processing
Dental Care
Desktop Publishing
Direct Marketing
Drafting
Electrical Trades
Electronics
Energy
Engineering
Engineering Technology
Environmental
Eye Care
Farming and Agriculture
Fashion
Fast Food
Federal Government
Film
Financial

Fire Protection Services
Fitness
Food Services
Foreign Language
Forestry
Franchising
Gerontology & Aging Services
Health & Medical
Heating, Ventilation, Air Conditioning, and Refrigeration
High Tech
Home Economics
Homecare Services
Horticulture
Hospital Administration
Hotel & Motel Management
Human Resource Management
Information Systems
Installation & Repair
Insurance
Interior Design & Decorating
International Business
Journalism
Laser Technology
Law
Law Enforcement & Criminal Justice
Library & Information Science
Machine Trades
Marine & Maritime
Marketing
Masonry
Medical Imaging
Medical Technology
Mental Health
Metalworking
Military
Modeling
Music
Nonprofit Organizations
Nursing
Nutrition
Occupational Therapy
Office Occupations
Paralegal
Paramedical
Part-time & Summer Jobs
Performing Arts
Petroleum
Pharmacy
Photography

Physical Therapy
Physician
Physician Assistant
Plastics
Plumbing & Pipe Fitting
Postal Service
Printing
Property Management
Psychology
Public Health
Public Relations
Publishing
Purchasing
Real Estate
Recreation & Leisure
Religious Service
Restaurant
Retailing
Robotics
Sales
Secretarial
Social Science
Social Work
Special Education
Speech-Language Pathology
Sports & Athletics
Sports Medicine
State & Local Government
Teaching
Teaching English to Speakers of Other Languages
Technical Writing & Communications
Telecommunications
Telemarketing
Television & Video
Theatrical Design & Production
Tool & Die
Transportation
Travel
Trucking
Veterinary Medicine
Visual Arts
Vocational & Technical
Warehousing
Waste Management
Welding
Word Processing
Writing
Your Own Service Business

VGM Career Horizons
a division of *NTC Publishing Group*
4255 West Touhy Avenue
Lincolnwood, Illinois 60646–1975